Conversations About Physics
Volume 1

Conversations About

PHYSICS

Volume 1

Edited by Howard Burton

Ideas Roadshow conversations present a wealth of candid insights from some of the world's leading experts, generated through a focused yet informal setting. They are explicitly designed to give non-specialists a uniquely accessible window into frontline research and scholarship that wouldn't otherwise be encountered through standard lectures and textbooks.

Over 100 Ideas Roadshow conversations have been held since our debut in 2012, covering a wide array of topics across the arts and sciences.

See www.ideasroadshow.com for a full listing of all titles.

ISBN: 978-1-77170-317-8 (hardcover)
ISBN: 978-1-77170-149-5 (paperback)
ISBN: 978-1-77170-148-8 (eBook)

Edited, with preface and all introductions written by Howard Burton.

All *Ideas Roadshow Conversations* use Canadian spelling.

Contents

TEXTUAL NOTE..8

PREFACE 9

INDIANA STEINHARDT
AND THE QUEST FOR QUASICRYSTALS
A CONVERSATION WITH PAUL STEINHARDT

Introduction... 17
I. Introducing Quasicrystals................................... 22
II. Building Models.. 29
III. Out of the Blue... 35
IV. Competing Explanations.................................. 38
V. Looking to Nature.. 50
VI. New Year's Delight... 56
VII. Confronting the Impossible........................... 60
VIII. Tracking Khatyrkite....................................... 64
IX. Kamchatka.. 77
X. Passing It On.. 83
Continuing the Conversation............................... 87

CRYPTOREALITY
A CONVERSATION WITH ARTUR EKERT

Introduction... 91
I. Beginnings... 97
II. Cryptographic Essentials................................ 108
III. Public Key Cryptosystems............................. 119
IV. Harnessing Interference 125
V. Quantum Sociology.. 139
VI. Quantum Metaphysics.................................... 149
VII. The Joy of Questioning................................. 162

THE PHYSICS OF BANJOS
A CONVERSATION WITH DAVID POLITZER

Introduction..171
I. The Feynman Experience...175
II. Love at First Sound 182
III. The Holy Grail 190
IV. The Ocarina Effect..195
V. Hearing Pitch...202
VI. Relative Strengths 208
VII. Transient Growth...217
VIII. The Working Physicist...226
IX. The Journey Continues...233

THE PROBLEMS OF PHYSICS, RECONSIDERED
A CONVERSATION WITH TONY LEGGETT

Introduction..241
I. Back to the Future...246
II. The Very Small...248
III. The Very Large...251
IV. A Glassy Digression..256
V. The Very Complex 259
VI. Understanding 264
VII. Different Regimes 267
VIII. Schrödinger's Cat...274
IX. The Slings and Arrows of Time................................279
X. The Anthropic Principle...286
XI. The Future of Physics...290
Continuing the Conversation..294

THE POWER OF PRINCIPLES
PHYSICS REVEALED
A CONVERSATION WITH NIMA ARKANI-HAMED

Introduction..297
I. Physics Time Management...304
II. The Problem with Popularization 308

III. In Feynman's Footsteps...311
IV. Describing Reality...315
V. A Timeless Community...321
VI. Against Relativism..324
VII. Strongly Constrained 328
VIII. In Search of a Formula 332
IX. A Principled Example 337
X. Supersymmetry...341
XI. Reacting Precipitously...346
XII. Tangled Pillars 351
XIII. The Pull of the Truth...354
XIV. Choosing a Better Description...357
XV. Beyond Space-Time..361

Textual Note

The contents of this book are based upon separate filmed conversations with Howard Burton and each of the five featured experts.

Paul Steinhardt is the Albert Einstein Professor in Science and Director of the Center for Theoretical Science at Princeton University. This conversation occurred on May 15, 2015.

Artur Ekert is Professor of Quantum Physics at the Mathematical Institute at the University of Oxford and Director of the Centre for Quantum Technologies and Lee Kong Chian Centennial Professor at the National University of Singapore. This conversation occurred on October 10, 2016.

Hugh David Politzer is the Richard Chace Tolman Professor of Theoretical Physics at the California Institute of Technology and co-recipient of the 2004 Nobel Prize in Physics. This conversation occurred on September 17, 2014.

Tony Leggett is John D. and Catherine T. MacArthur Professor Emeritus of Physics at the University of Illinois and co-recipient of the 2003 Nobel Prize in Physics. This conversation occurred on August 8, 2013.

Nima Arkani-Hamed is Professor of Natural Sciences at the Institute for Advanced Studies in Princeton, New Jersey, where he pursues his wide-ranging interests in fundamental physics. This conversation occurred on November 12, 2012.

Howard Burton is the creator and host of Ideas Roadshow and was Founding Executive Director of Perimeter Institute for Theoretical Physics.

Preface

When I was running a theoretical physics institute, by far the most common question I would be asked by people on the street was simply, *"What do those guys actually do all day long?"* It's the sort of thing that I make a conscious effort to bear in mind as Ideas Road-show conversations meander towards intriguing ruminations on the thermodynamic arrow of time or physics beyond the Standard Model. Detailed, high-level speculations are all well and good, but it's important not to lose sight of what "thinking like a physicist" is really all about: What are the common approaches? How can progress be made when there is no experiment to guide us? To what extent can we even be certain that "progress" is being made at all?

Good questions, all. In this collection the motivated reader will find a spectrum of different responses from some of the world's leading theorists—some similar, some distinct, but invariably deeply intriguing and thought-provoking.

One particularly revealing portrayal of the problem-solving approach of a theoretical physicist occurs in the musings of Princeton University physicist **Paul Steinhardt** as he tried to ponder the meaning of an unexpected symmetry that appeared in his computer simulations of amorphous metals when he was a young postdoctoral researcher.

> *"We were using simulations that just had a single type of atom in them, and I was interested in whether one could get this icosahedral order to go a little bit further. Let's say we had two types of atoms—because real amorphous metals, for example, often have two or three types of atoms in them—I thought that maybe that would enable a more extensive degree of icosahedrality. Now most sane people would have said, 'Well, maybe, but you're only going to get a little bit*

more, because this is one of the most famous forbidden symmetries for solids.'

"The icosahedron has five-fold symmetry axes along every pair of vertices—so that makes six of them, since there are 12 vertices in total. It's the most famous forbidden symmetry for solids. At that time, if you were to open up any elementary solid-state physics text book, usually within the first chapter there would be a picture of an icosahedron explaining that you will never get this symmetry for a solid.

"But I wanted to know how we really know that. I agreed that it couldn't be a crystal—but of course that wasn't what we were making in our simulations. It wasn't crystalline, but on the other hand, it did have this intermediate-range icosahedral order. Who says that can't be extended?" What would that look like? How might it be possible?"

And then there's University of Oxford and NUS quantum information theorist **Artur Ekert**, who famously found a way to harness the inherent weirdness of the quantum world as initially framed by Einstein and his colleagues in the 1930s, concretely reformulated by John Bell in the 1960s, and experimentally demonstrated by Alain Aspect in the 1980s.

"I was simply driven by philosophical curiosity, saying to myself, 'Well, if something doesn't exist, there's no element of reality; and if information is physical and thus encoded in something, and you've encoded it in something that is, somehow, losing its element of reality, then there's nothing to eavesdrop, because it doesn't exist.' So that was my thinking. And this demonstrates that quite often progress is not driven so much by rigorous mathematics, it's driven by these other crazy things, like 'an element of reality'.

"In his EPR paper, Einstein defines this element of reality very clearly. He says, 'If in any way without disturbing the system, you can learn about the value of a physical quantity, then you attribute an element of physical reality to it.' So the notion of "disturbance" was clearly specified there, and I thought to myself, Well, this is exactly the definition of eavesdropping in cryptography. To me, the definition

of the "element of reality" in the EPR paper was reading like a definition of perfect eavesdropping.

"And then I thought, But wait a minute, now we know something about the non-existence of this element of reality thanks to Alain Aspect and John Bell, therefore if I exclude the existence of the element of reality, I exclude the existence of perfect eavesdropping. That was my independent way of thinking about designing a quantum system that would make perfect eavesdropping impossible—or, you could say, any attempt at eavesdropping detectable."

Caltech physicist **David Politzer**, meanwhile, who was awarded a Nobel Prize for his work in particle physics that he did while a graduate student, describes how he later became attracted to a problem worth addressing in solid state physics.

"I was intrigued by one simple fact, which is kind of marvellous: it's a feature of the physics of the very small. Heat in a solid is represented by vibrations—we call them phonons—the quanta of vibrations, which have wavelengths. When it gets cold, those wavelengths get longer and longer. You can make something small enough, and it's got billions and billions of atoms and it can be small enough and cold enough such that the wavelength of the relevant heat quanta is much bigger than the thing.

"So it's essentially at zero temperature, because it has no thermal motion. It's in contact with things that have a temperature, and it's in equilibrium with it as a temperature, but there's no motion there. So you're now at zero temperature, and you can then tease apart—and this became a huge field of condensed matter physics—the thermal physics from the quantum physics.

"There were questions there that I thought we should be able to answer. Let's put it that way. It wasn't hugely speculative. There was a way to carve out some very straightforward, Politzer-like, question that would have an answer; and some of it I was able to answer. That made me feel good."

Our final two examples illustrate the sort of thought experiments that theoretical physicists often conduct to help them puzzle their way towards a deeper understanding in the absence of any sort of guideposts from experiment, computer simulations, or the empirical world in general.

Curiously enough, as it happens, both of them involve appreciating historical parallels to our level of ignorance through hypothetical time travellers.

Here's Nobel Laureate **Tony Leggett** musing over the possibility that the laws of quantum theory might break down at some future scale factor in a very similar way to how they mysteriously arose in the first place.

> *"Imagine that I'm a time traveller travelling back from today to the year 1870 or thereabouts. I meet a group of physicists and I assure them that, at some point, as you go down in scale from the order of the macroscopic world to the atomic world, the laws of mechanics are going to change radically and fundamentally.*

> *"They would look at me, somewhat perplexed, and say: 'OK, but what's going to change? After all, mechanics has no scale in it.' And I would say: 'Well, sorry mate. Actually, according to modern information there is a scale.' It would not have been reasonable for them to guess it at the time, but it is there.*

> *"I would speculate—and in some sense hope—that even after the next major revolution, if there is one in physics, we'll still be able to use quantum mechanics for the purposes of describing phenomena at the atomic level, just as we now still use Newtonian mechanics to describe planetary motion. Nothing has gone wrong with that."*

And here is Institute for Advanced Study particle physicist **Nima Arkani-Hamed** invoking his own "time travel thought experiment" to argue that we need to find a new way of rephrasing our current expression of the fundamental laws of physics that will meaningfully direct us to a deeper level of understanding.

"You have to have something to do: you have to have some concrete angle into the problem. The most obvious angle into difficult problems is experiment, but we don't have that for this question of where space-time comes from. We're not going to do experiments at the Planck scale any time soon. So my thinking is to look at the past for a guide. We've been through a similar huge conceptual leap before, when we went from classical to quantum mechanics. It's hard to imagine a bigger leap than that: we thought the world was deterministic, but it turns out that it's not.

"And I imagine what it would be like if you're a classical physicist in the year 1790 or something. You're working away and using Newton's laws to solve technical details related to the orbits of planets and stuff like that. And suddenly you're visited in the middle of the night by the Ghost of Theorist Future, who announces, 'I have a message for you from the year 1930: determinism is gone.' And then he disappears into the night, as Ghosts of Theorist Future are wont to do.

"Now obviously, you're very excited by this. What are you going to do with this information? How does it change what you're doing? How, specifically, are you going to incorporate this news? You obviously want to head towards this deep important thing that's going to come centuries from now.

"And there is something that you can do, which is to say, 'Look, if there's really no determinism, then there has to be some way of talking about even the physics I have right now under my feet in a way that somehow doesn't put in determinism. It can't really be there, so there must be some way of talking about it that doesn't have determinism in it.' And that's a very startling thing. Because now, you see, you're on a straight and narrow path because you're not trying to guess the right answer, you're not trying to just mess with things randomly—instead you're trying to take this clue to force you to think about what you have under your feet in a radically different way.

"And it turns out there is another way of thinking about good, old-fashioned classical physics: the idea that uses the principle of least action. As it happens, the people who discovered the principle of least action were really startled by it philosophically, precisely

because it looked so unlike Newton's laws. It didn't look deterministic, but, of course, it was just a rewriting in the end: it was equivalent to Newton's laws. We now understand why it is that the principle of least action exists. It's because of quantum mechanics! The world is actually quantum mechanical, it isn't deterministic. And, indeed, this basic philosophy is correct in the limit as you can ignore quantum mechanical effects: the quantum formulation of the physics reduces to this new way of thinking about classical physics, which is the principle of least action."

Understanding the precise theoretical frameworks that physicists have developed to attack today's fundamental problems is admittedly very difficult. Even appreciating what those problems are can be maddeningly complex. But getting a sense of how, in general, they are going about doing so turns out to be remarkably accessible—and deeply captivating.

Indiana Steinhardt

And the Quest for Quasicrystals

A conversation with Paul Steinhardt

Introduction

Informed Authority

At some point during my Ideas Roadshow conversation with the renowned polymath Freeman Dyson, I asked him to compare today's crop of physicists to those great titans of the past, many of whom he had worked closely with. As I expected, he was politely respectful of his current colleagues but hardly evinced a great deal of genuine enthusiasm, citing the distinct lack of original, independent thinkers—"heretics" he fondly referred to them—at one point frankly admitting, that *"Physics is not in good shape at the moment".*

Yet one clear exception to Freeman's general sense of physics ennui was Paul Steinhardt, a name he kept coming back to with eyes shining, so much so that at one point I felt compelled to dryly remark, *"This seems to be turning into the Paul Steinhardt show."*

Freeman, however, didn't flinch.

"He's one of the real heroes," he told me earnestly, adamantly maintaining that I simply *must* talk to Paul about his quasicrystal adventure that culminated in a dramatic Siberian expedition. *"That's an example of physics that is still going strong,"* he enthused.

As it happens, I knew Paul a bit already. For some years he had been the Chair of the Scientific Advisory Committee of the research institute I was in charge of, so we had spoken quite often. But interacting with someone in a purely administrative manner—particularly when an important aspect of his job is to critically assess what you are doing—naturally tends to colour the nature of the relationship.

I was aware, of course, that he was a highly accomplished scientist who had made significant contributions to both cosmology and condensed matter physics, and through our many conversations I also learned that he was an amiable and personal fellow along with being a die-hard New York Yankees fan. But none of that, naturally, prepared me for Freeman's outburst.

It should also be said that, like most people trained in the scientific tradition, I have a knee-jerk, almost childish, antipathy to anything smacking of an appeal to authority. So when I'm assured that someone is unequivocally wonderful—even if the assurer is himself as unequivocally wonderful as Freeman Dyson—I am naturally somewhat sceptical. But I did make a mental note to talk to Paul about this Siberian quasicrystal business, which certainly seemed intriguingly different. I was already hoping to talk to him about his iconoclastic views on inflationary cosmology—a field he had strongly contributed to developing and was now sharply criticizing—and thought it likely wouldn't be too difficult to add a few minutes of quasicrystals to the mix during our discussion.

Well, it turned out to require more than a few minutes. In fact, it turned out to require an entirely different conversation.

I knew a tiny bit about quasicrystals before we met—that they were a new form of matter that was sort of crystalline and somehow mathematically related to Penrose tiles—and that Paul had something to do with their development. That was it.

It should probably also be mentioned that, growing up, I had never been terribly interested in geology, crystallography, the physics of materials, or any of that. Even solid-state physics seemed a bit too applied and boringly contingent for my tastes, as, like many students unthinkingly swallowing the "secrets of the universe" cool-aid, I set my sights on what I believed were deeper, more "fundamental" things.

By the time I unexpectedly found myself running a theoretical physics institute, however, my views had evolved somewhat. I began to appreciate that an impressive proportion of some of the greatest minds in physics—whether spurred on by a desire to probe the precise nature of our theoretical understanding or just because they were intrigued by a scientific puzzle—were captivated by the question of just how, exactly, materials around us are put together.

Because nature, it seems, is continually surprising us. And just about every time we think we know what's going on, it turns out that we don't.

Paul Steinhardt figured this out long ago. As a young post-doctoral researcher, he developed a computer model to probe the structure of so-called "amorphous metals"—metallic alloys that are cooled in such a way that their underlying chemical structure becomes random-like —like glass—rather than ordered in a crystalline pattern.

The goal was to see to what extent the final structure of these amorphous metals was truly random. It turned out that there was an unexpected type of ordering induced in the atoms of these materials—a short-term correlation that took the form of a soccer ball-like icosahedron. Interesting, perhaps, but hardly what you might think would be the start of a revolutionary tale.

Well, the thing about surprises is that you simply never know what might happen next. For Paul, it turned out that the key question involved investigating the possible limits to what his computer model had found.

> *"It wasn't crystalline, but on the other hand, it had this intermediate-range icosahedral order. **Who says** that can't be extended? **What** would that look like? **How** might it be possible?"*

From there, spurred on by the mathematical insights of Roger Penrose's so-called "quasiperiodic" tilings, Paul and his students began to develop a framework for what new states of matter might, in fact, be possible—even going so far as to demonstrate how such so-called "quasicrystals," if somehow created, might actually be detected.

Shortly afterwards, as seems to happen not infrequently in science, precisely such a material *was* serendipitously created by material scientists—eventually earning Dan Shechtmann the 2011 Nobel Prize in Chemistry—thereby demonstrating that quasicrystals weren't just mathematically possible, but could actually be made to exist in the real world. Nature surprised us again, it seems.

Once it was clear that quasicrystals could be synthesized in a laboratory, it was time to start asking whether or not nature had bothered to produce them herself.

And here is where our tale leaves the realm of computer simulations, mathematical equations and laboratories, and rushes headlong into nothing less than "Indiana Jones" territory.

In the pages to follow you'll encounter mineral smugglers, secret diaries and quasi-mythical characters in a tale that tacks from Florence to Israel, Amsterdam to California, Princeton to Kamchatka. And at the end of it all, you'll be presented with something that is, quite literally, out of this world.

Perhaps even more significantly still, you'll get a clear sense of all that science can be.

Sometimes, it seems, appeals to authority are worth making. Because Freeman Dyson, as usual, was right on the money.

The Conversation

I. Introducing Quasicrystals

Innovative symmetries through visual disharmonies

HB: Let's get right to it by starting with the obvious questions: what is a quasicrystal, and how did you get involved in the whole business?

PS: Quasicrystals are a new form—a new phase—of matter, which violates the rules of symmetry that we thought all solids had to obey. When atoms or molecules come together to form a solid, they arrange themselves in some sort of three-dimensional configuration. They pack together.

Before the discovery of quasicrystals, we knew of two general kinds of organization. One was random. For example, if you take a liquid and cool it very rapidly, it will freeze into a random arrangement of atoms or molecules. Ordinary glass is like that, only a little bit more complicated. It's actually a frozen network—the atoms are bonded in a kind of network that, while randomly arranged, is twisted and turned and sometimes has vacancies and holes in it.

The other kind of arrangement, which we often find aesthetically pleasing, is the crystal arrangement. That—or so we thought—was the lowest energy arrangement possible for atoms and molecules. In crystals, the atoms and molecules arrange themselves like building blocks in a child's toy. A building block would represent a certain arrangement of atoms, and then the way the solid is built up is by just repeating that arrangement over and over again.

Sometimes the building block is just one atom—like in a copper metal. Sometimes the arrangement can be quite complicated—like DNA—but it still has this property that it repeats over and over again. In some ways, it's a little bit like a tiling in two dimensions—if you were trying to tile your bathroom floor with only one type of tile, you would want to find the single type of tile which would fill the entire plane—in this case, the two dimensions of your bathroom floor.

What we learned about 200 years ago is that there are strict rules about what symmetries or forms you can have. Actually, we discovered it empirically long before then—probably at least as far back as the ancient Egyptians. They already explored all the known patterns, but in terms of mathematics, we didn't understand it until the 19th century.

Let's imagine that we're going to tile our bathroom floor and we're going to use identical tiles. You want to decide what patterns you can make. You could imagine doing it with squares, triangles, rectangles, parallelograms, hexagons—you might think the list would go on and on, but that's actually the end of the list. You can't do it with any other regular shape. If you tried to tile your floor with, say, pentagons, you'd be pretty upset because you'd find that there are regions that can't be filled in with those pentagons: there would be holes.

You've probably seen patterns with octagons, but if you think about it carefully, it's not just octagons; it's octagons with little squares in the corners. And that is really a square pattern in which you've inscribed an octagon inside each one.

Because we get bored with this small set of patterns, humans have figured out all kinds of ways of hiding that order and symmetry. The great 20th-century Dutch artist M.C. Escher, for example, was a marvel at doing that kind of thing in his drawings. But still, there's this strict restriction—or so we thought—for both patterns and matter.

Quasicrystals shatter that picture because they have symmetries that are forbidden, symmetries which are not on that list. Quasicrystals can have the symmetry of a pentagon, or seven-fold, or eleven-fold symmetry, as well as other similarly forbidden symmetries in higher dimensions.

That's because they introduce one new element into the story, which opens up the symmetry possibilities.

Whereas in crystals, the atoms and molecules repeat at a single frequency; for quasicrystals, they repeat at two or more frequencies where the ratio of the frequencies is irrational—not expressible as a ratio of integers. What, in sound, we'd call a disharmony.

If you heard a sound that was a combination of two tones which were disharmonic, your first reaction would probably be that it's just noise. But if you took its Fourier transform, if you analyzed its tones, you'd actually find that it's *not* noise.

Noise would have a broad spectrum of frequencies, but here you just have two tones, which happen to be in disharmony. In other words, it's actually a very special, ordered sound—it's just one that your ear has trouble interpreting and distinguishing from noise.

Visually, quasicrystal patterns have that feel to them too. When you look at them, depending on how you view them, you might first say, *"I'm looking at some kind of ordered pattern,"* and then try to figure out what's going to happen next in that pattern. And you'd find that it's kind of unpredictable, so you might think it was random. In other words, without any training, humans cannot instantaneously recognize a quasicrystal from something that is random.

So it's hidden in the sense that, unless you knew how to analyze it properly, you wouldn't realize what you had.

The idea of quasicrystals was first introduced by my student Dov Levine and me in the beginning of 1981 as a way to break the established symmetry rules. We were interested in making solids that would violate the usual rules of crystallography.

HB: How did you come to start thinking about this?

PS: Well, like most of my stories about science, it's going to be one with lots of twists and turns. If we go way back, it probably began when I was a kid, beginning to read about group theory and crystallography. I just thought it was an amazing thing that, with pure mathematics, you could predict all the allowed shapes of crystals.

It's actually considered one of the great triumphs of theoretical physics that you could completely categorize all the possible shapes just by doing a mathematical calculation having to do with symmetries and the mathematical properties of symmetries. I was just really struck by that, since I was interested in pattern-making and analyzing patterns. Even today, if I'm in a room or some place and there is some sort of pattern, I'm going to be looking at it and, for whatever reason, trying to analyze it.

So that was, I suppose, always in the back of my mind. And then when I was an undergraduate at Caltech, I was going through this process of trying to figure out what I was going to do in my life. I decided it would be physics, but I still didn't know what specific area of physics.

I spent one summer in a program at Yale University being introduced to condensed matter physics—the physics of solids. In particular, I was tasked with trying to make what, at the time, was the first computer model of amorphous silicon.

People had made physical models of it that were supposed to represent the random network arrangement of silicon. If you make crystalline silicon, like you have in many of your transistors, that makes a beautiful "diamond lattice" of hexagons. But if you cool the silicon rapidly, it forms amorphous silicon in which, instead of getting all six-fold rings, you sometimes have five-fold and seven-fold rings—it forms a twisted, complicated network.

People had made physical models, but my summer job was to try to build a computer model that we could then study in detail to measure lots of properties. I just found it fascinating that if you make it crystalline we understand it completely, but if you make it non-crystalline we don't really understand it very well at all. Actually, we don't even completely understand it today. There are still some significant, open issues about how to properly understand it.

Then zoom forward a few years, and I was invited to IBM Research to analyze, not silicon, but amorphous metals. Amorphous metals are a little bit different—they don't form a network. Metals pack like spheres, so it's like throwing a bunch of marbles together randomly and asking what properties they have and what the right way to characterize their arrangement is: is it *truly* random, or is there some hidden order in there?

I was then a postdoc working with David Nelson, a condensed matter physicist at Harvard. He had the idea that, if you were to study simulations of these amorphous metals carefully, you would find that there's a hidden, what we call, cubatic order—if you look at the two nearest neighbour atoms and the line segment between them and then compare them to another pair, that although there is some degree of randomness, on average they would tend to be aligned along a cube. In other words, they're not strictly positioned cubically, but the alignment has a kind of cubic sense to it.

At that time I think I actually had the world's largest computer simulation of amorphous metals, so I had the right tool for the job. We began to analyze the simulations and we didn't find this cubatic order, but we found something else instead—we found icosahedral order, which is the symmetry of a soccer ball.

An icosahedron is one of the five Platonic solids. The name "icosahedron" comes from Greek, literally meaning "20 faces" since it has 20 triangular faces, all of which are identical, along with 12 identical vertices and 30 identical edges or links.

So this was the local symmetry, in the sense that a link in one place would be, on average, oriented similarly to a link in another place. It wasn't an infinite range icosahedral order: it was like this over a fairly substantial size of the computer simulation, but after about 10 interatomic spacings or so it began to peter out.

There was, then, a local degree of icosahedrality, which was unexpected, but I think, at least in David's view, that was kind of the end of the story. That was the best you could do.

We were using simulations that just had a single type of atom in them, and I was interested in whether one could get this icosahedral order to go a little bit further. Let's say we had two types of atoms—because real amorphous metals, for example, often have two or three types of atoms in them—I thought that maybe that would enable a more extensive degree of icosahedrality.

Now most *sane* people would have said, "*Well, maybe, but you're only going to get a little bit more, because this is one of the most famous forbidden symmetries for solids.*"

The icosahedron has five-fold symmetry axes along every pair of vertices—so that makes six of them, since there are 12 vertices in total. It's the most famous forbidden symmetry for solids. At that time, if you were to open up any elementary solid-state physics text book, usually within the first chapter there would be a picture of an icosahedron explaining that you will *never* get this symmetry for a solid.

But I wanted to know how we really know that. I agreed that it couldn't be a crystal—but of course that wasn't what we were making in our simulations. It wasn't crystalline, but on the other hand, it did have this intermediate-range icosahedral order. *Who says* that can't be extended? *What* would that look like? *How* might it be possible?

Shortly after that I moved to the University of Pennsylvania. Dov Levine joined me as a student and he foolishly decided to join the project, even though I told him that it was a really impossible project and was unlikely to lead anywhere. Finally I said to him, "*Well, if you insist on joining this crazy project*"—which he did—"*let's get started.*"

We began by trying to build icosahedral things. We built icosahedra, icosahedra of icosahedra, trying to determine what happens: how, exactly, we run into trouble. Then at some point Dov brought in a *Scientific American* article with a picture on the cover of a Penrose tiling and said, "*Maybe we should be thinking about this.*"

Questions for Discussion:

1. Did you find Paul's analogy of two disharmonic tones helpful? To what extent do analogies such as this one help us to focus on core conceptual issues of a problem?

2. In what ways does this chapter illustrate the interaction between theoretical frameworks, computer simulations and experiment? To what extent do you think Paul's computer simulations needed regular input from those working with the materials in the lab?

II. Building Models

Forcing forbidden symmetries

PS: Penrose's tiling is a nonperiodic tiling of just two tile shapes. If you look at the pattern, it has local arrangements which look five-fold, so it suggests a kind of five-fold symmetry, which would be crystallographically forbidden.

Penrose was trying to make something nonperiodic—which says what it *isn't*, but it didn't say what it *is*, or what the secret behind it was. So that's what we got interested in, and we wanted to understand it better.

We wanted to know, first of all, if it really *was* five-fold symmetric. We eventually showed that it was, because every link in the system, every edge of the tile, is always oriented along one of the edges of a pentagon; and each direction is represented equally. So that means it really *does* have a statistical, average, five-fold symmetry.

Then we asked ourselves, *OK, so if it's **not** periodic, is it anything else?* And what we discovered is that it's actually quite special. You have these two different tiles, and each one repeats with a different average frequency where the ratio of those frequencies is an irrational number. So it's what's called "quasiperiodic".

The secret, apparently, for getting that symmetry, which was forbidden, was the fact that it was quasiperiodic rather than periodic. The 200-year-old theorems of crystallography always said, *If you have a crystal where things are periodic, then here are the symmetries that are allowed.*

And people had only imagined one other possibility: random. But now we were imagining this two-tone-like possibility, to use the analogy I made earlier, which suddenly enabled you to get *new* symmetries. In fact, you discover that all of the symmetries that we thought were forbidden can now be achieved using quasiperiodicity.

HB: Could you explain why the ratio between the two has to be an irrational number?

PS: Well, if it would be a rational number, then it's like making two tones that are harmonic, so the tone of one will be in phase with the other every now and then, which means that on that scale, the sound is periodic: it's repeating. In space, it would mean that, every now and then, you'd be in sync again, and that "in-syncness" would repeat again and again. In other words it becomes periodic, just on a larger scale. But that wouldn't happen if the ratio between the two tones was an irrational number.

And that irrational number is set by the symmetry. In other words, if it were rational, you couldn't get the five-fold symmetry. Five-fold symmetry demands irrationality. It even tells you what the irrational number has to be. It's not a random irrational number: it has to be the golden ratio—$(1+\sqrt{5})/2$—and that's fixed by the geometry, by the sines and cosines of angles associated with pentagons or icosahedra.

The same is true with icosahedral symmetry. If you allow me quasiperiodicity, I can also make something which is icosahedrally symmetric—the most famous forbidden one—which was the kind of symmetry we were aiming for. So we knew that we could make something icosahedral this way.

Penrose's tiling had another feature that was very important, which wasn't simply that it could be made nonperiodic, but if you label the tiles appropriately, or put jigsaw-puzzle-like holes and hooks into each one, you could have two types, which could *only* fit together to form the Penrose tiling—which he called nonperiodic, but now we would say it's quasiperiodic or quasicrystalline.

The question was, *Could we do that for the icosahedral case?* It wasn't clear that you could do that, but it was important to us because what it represented was that, not only was it possible, but you could even imagine atomic interactions which would mimic, or have the same effect, as those little jigsaw puzzle pieces—preferring some arrangements over others. In which case, the quasicrystal would be energetically preferred, maybe even more preferred than a crystal.

So we spent some time trying to do that, and we thought it was going to be a simple thing to do, but it turned out to actually be quite complicated. In the end, we found a couple of schemes. I found a scheme for doing it with Dov, and then another scheme I found with another student, Josh Socolar.

I prefer the scheme that Josh and I developed, because it has other properties that I find interesting.

So now we're dealing with three dimensions and it turns out, for the icosahedral case, you need four types of units if you want to have forcible tiles. If you just want to fill the space in a quasiperiodic pattern, you can do it with two types. But if you want to have units which can *only* fit together to form it, you can't do that with two: you need four, as far as we know.

This is one example of the choice of four different units.

They're actually a family of shapes known as zonohedra. They're projections of different pieces of hypercubes from six dimensions into three dimensions.

I didn't say it before, but if you have something which is quasi-periodic, one of the things you immediately know is that you can always view it as a projection, or a cut, through a crystal lattice of higher dimension. In the case of the icosahedral symmetry, the fact that it has six 5-fold symmetry axes ends up dictating that you need to go to six dimensions. Then, with an appropriate cut through that cubic lattice in six dimensions, you can get the appropriate tiles by a projection into three dimensions.

For example, this is the projection of a six-dimensional cube in three dimensions.

HB: So these are fascinating models, but I'm thinking, Okay, he's playing all these mathematical games. That's interesting, but does it have anything whatsoever to do with anything in the real world?

PS: Well, that's the question we wanted to answer.

So we now had the idea that you *could* have forbidden symmetries, and you could even force them through—in this case Lego-like interactions—but if you can do it for Lego-like interactions, in principle, there might exist *real* materials that would have them.

We had even computed what the diffraction pattern from these materials would be, because we wanted to know, *If you **were** able to find such a material in the laboratory, how would you even **know** that you had it?*

Going back to Von Laue and Bragg in the 1920s, we knew that the way of determining the structure of a crystal is to scatter X-rays, or electrons, or neutrons through it—they go through the channels of the material, come out the other end and interfere with one another, much like the way light interferes with itself as it goes through a set of slits producing a pattern which we call a diffraction pattern.

For crystals that pattern is very distinctive: it's a lattice of pinpoints—they're pinpoints because the structure is periodic. The lattice displays a symmetry associated with the atomic arrangement along a given direction, which might be different in different directions. There would be a basic spacing between the atoms, and that basic spacing would have to do with the size of the repeating unit. We use these basic rules of diffraction to solve the atomic structure of many materials: minerals, synthetic materials, DNA, and so forth. This is standard practice.

If you had something in which the atoms were randomly arranged, you wouldn't get pinpoints at all; you would just get a sort of cloud coming out the other end—something that's just diffuse: symmetrical, and round.

For the quasicrystal, what we found is that, if you have a structure which is quasiperiodic, that means it's really made up of periodic elements, each of which would produce pinpoint diffraction. So you're *going* to get pinpoint diffraction, but the symmetry of the lattice of pinpoints is going to be *different*, first of all because it's going to have a symmetry that is not one of the symmetries you were supposed to have.

For example, if you were going to do this for the Penrose tiling, you would get something where the symmetry is ten-fold—which is one of the forbidden symmetries—or if you were going to do it for the icosahedral lattice, looking along one of the five-fold symmetry directions, you'd get something similar.

So the arrangement of the pinpoints, just the symmetry of it, is very different; and secondly, because of the quasiperiodicity, you have more pinpoints as you go along any direction in your pattern rather than having a single fundamental distance between pinpoints.

You could think of it as beginning with two different distances which are irrationally related, plus all combinations of them, plus and minus. And if you make combinations of all pluses and minuses, between every pair of pinpoints, there should be yet more pinpoints, and yet more pinpoints, and so on.

In the end, then, what you get is not something that looks like a crystal lattice with the basic spacing, but rather a kind of snowflake pattern, a self-similar pattern of pinpoints. It's very distinctive. So we knew what to look for if you were to find such a material.

Questions for Discussion:

1. Why do you think the notion of symmetry has played such a strong role in our understanding of nature?

2. What, exactly, do you think that Paul means when he says when describing the situation of two tones repeating with a ratio that is a rational number, "In other words, it becomes periodic, just on a larger scale"?

3. Are there some concepts about the physical world that are fundamentally impossible to comprehend without a sufficient understanding of mathematics?

III. Out of the Blue

The real world intervenes

PS: Now, *how* might you find such a material? Well, by this point, I thought we had gone far enough along with this idea that I would take a leave to go to IBM Research in Yorktown Heights and try to convince people there to try to either fabricate or look for quasicrystals. I had some ideas about how to look for them.

HB: Looking for them, I understand. You've done the diffraction analysis, so you have some characteristic signature that you can identify. But how do you go about trying to fabricate them?

PS: Right. In terms of looking for them, one thing you can do is just conduct an exhaustive search through lots of unusual materials. That's hard to do, but you could do it that way.

But the fabrication idea that I had in mind at the time involved using colloids.

You could think of colloids as condensed states made, for example, out of little beads of liquid, where the beads are charged so they have forces between them. It was known that they could arrange themselves in crystal patterns, and I had already sort of worked out a scheme for how I might be able to get it to work for a quasicrystal pattern.

HB: Tell me more about this scheme. I know nothing about this stuff and I'm wondering how you actually move these guys around.

PS: Well, they have to organize themselves. So you could use beads with two different charges on them, which would induce two different forces, two different characteristic lengths: that was the idea. I tried to play with tuning those charges and tuning the ionization of the fluid they were in. You're pretty limited in the degrees of freedom you can play with, so it's not clear that you can do it. And I don't believe anyone has done it to this day for icosahedral symmetry. But that's what we were trying to do.

There were groups there that worked on colloids and characterizing crystals, and I had a long and productive relationship with IBM, so I went there on leave with the intention of doing that.

But shortly after I arrived, Dave Nelson—the fellow I mentioned earlier with whom I had done a postdoc years before that had gotten me started in this whole direction—came to visit to give a colloquium at IBM. I was very excited to see him because we hadn't talked for a few years and I wanted to show him all of this great stuff we had done on quasicrystals in the meantime.

He came by my office. I thought he was coming to hear what I had to say, but instead he said, "*Actually, I want to show **you** something.*" And I said, "*Well, I want to show **you** something.*" We went back and forth like this for a bit, but eventually I let him go first. He pulled out a preprint and said, "*I think **this** is something you might be interested in.*"

It was a preprint by four people—Dan Shechtman, Ilan Blech, Denis Gratias, and John Cahn—who were working at the National Bureau of Standards, as it was called at that time. They had found a material that had a very unusual diffraction pattern. They had something that was icosahedral, but they didn't understand what it was.

I thought, at first, that what he was showing me was something like what he and I had worked on—something with short-range order—but as I began to read, I realized that they were talking about something with *long-range*, icosahedral order. Then I flipped the page and there was the same diffraction pattern I had been working on. I think I jumped out of my chair.

I don't think I said anything; I just walked over to my desk and picked up the diffraction pattern that I was going to show *him*—of the model I just showed you—and I just put it next to the other one from the preprint and said, "*I think we have the answer to this question.*"

Dov was visiting at the time—because I wanted him to meet with Dave when we explained our ideas to him—and of course he immediately recognized what was going on. That was the *"Aha! moment,"* where we knew that this theoretical concept actually had some reality: someone had discovered such a material by accident, which was incredibly exciting.

Questions for Discussion:

1. *What exactly do you think Paul means when he says, "You're pretty limited in the degrees of freedom you can play with, so it's not clear that you can do it"?*

2. *Does it surprise you, based upon what you've read so far, that Paul was not a co-recipient of the 2011 Nobel Prize that was awarded to Dan Shechtman for the discovery of quasicrystals?*

IV. Competing Explanations

A three-horse race

HB: So then what did you do? Did you phone these guys?

PS: Pretty soon afterwards I phoned people at Penn, where I was at the time, because they had been supporting all this work over the years, even when people thought it was crazy to be working on these things.

It was very abstract stuff—to be working on hypothetical forms of matter—so I wanted to let them know that it actually may be very interesting and relevant. People there immediately got to work on trying to reproduce the results in that paper, and they made many important contributions.

I also contacted the authors of that paper. John Cahn came up and we had a session with him. Then, of course, we wrote up our results and submitted them; and a few months later there was a grand presentation at Penn where John came up and we presented our theory and they talked about their experiment. That was kind of how it rolled out to begin with.

It was a nice beginning, but like often happens when experiments produce unexpected results, there begins to be some controversy about different ideas. We had an idea of what might explain their material, but other people, looking at the same information, had competing ideas.

HB: What were those ideas?

PS: Well, there were two broad categories of ideas.

One was that, whereas the picture *we* had was that these were really ordered materials in the sense that they produced pinpoint diffraction—they're just quasiperiodic rather than periodic—a competing idea was that they were just a random arrangement of icosahedral clusters distributed spatially with no long-range order at all.

It turns out that if you make such an arrangement it makes a diffraction pattern which looks very similar to Shechtman's, but because it's disordered, you don't get pinpoints, and instead have a kind of broad spot. For this model, the spots would be in the right place on average, although they might not be if you just looked at a particular instance.

In fact, Shechtman's material really wasn't that good. It was fair to say that this model explained the data as well as ours, because his pattern really wasn't truly pinpoint and the spots weren't exactly in the right positions.

From our point of view, we'd say, "*Well, what do you expect? The way you made your material was by rapidly cooling it. It should have lots of defects, which will produce that diffraction effect.*"

But one could also take the point of view, "*No, actually, the quasi-crystal picture isn't right. What we have here is inherently a random arrangement with no long-range order.*"

So, there was that kind of difference between what we call the quasicrystal and icosahedral glass views—with "glass" here implying this randomness.

Meanwhile the third idea was one that Linus Pauling was promoting.

HB: Did this have something to do with vitamin C?

PS: No, that's not what he was promoting in this instance.

He wasn't a fan of quasicrystals—indeed, he was reputed to have said something like, "*There is no such thing as quasicrystals, only quasi-scientists.*"

At any rate, he had a crystal model that he thought could explain the data just as well. The reason why you could invoke a crystal model here was because the points weren't quite on the right spots in Shechtman's material, and if you deviate ever so slightly then what you thought were irrational ratios can be fit well by rational ratios that could then give you a crystalline pattern.

His idea was that the phenomenon in question was actually "multiply-twinned" crystals: crystals rotated by certain angles with respect to one another to mimic something icosahedral, even though it's inherently crystalline.

It's not a crazy idea. Many materials grow that way. If you grow small particles of gold, they will form an arrangement that will look very much like the icosahedral arrangement. But if you look carefully into the interior of each face of the icosahedron there will be a tetrahedron of atoms that will all be arranged like a crystal. In other words, what you have are 20 separate grains, each of which is crystalline, but they happen to arrange themselves in this one, large configuration like an icosahedron, this particular nugget.

Two grains of crystal that are the same but differ in their orientation are called "twins." In this case, we would have a "multiply-twinned" crystal, because we wouldn't have just 2 identical grains oriented differently, but 20 identical grains to make up the icosahedral nugget.

So it wasn't a crazy idea at all. It was a fairly sensible idea, actually.

HB: You're pointing out all these different structures and patterns that can arise in nature, but a key question for me—as someone who knows very little about this—is, *How does this happen to begin with? How does nature actually create these sorts of patterns, anyway?* Maybe it's too difficult to explain here.

PS: Yes, it is difficult. For simple patterns—those with just a few atoms involved, maybe one or two types of atoms, a very simple arrangement—we have pretty good arguments and calculations that can explain why that particular structure is energetically preferred, enabling you to understand why it would be *that* crystal, say, rather than something random or something else entirely. But for the complicated structures we're talking about, we actually *don't* have that kind of power in our understanding of materials.

For example, if you tell me, "*I'm going to mix three random elements in the Periodic Table*," and I haven't done it before so I don't know the answer, there's no way I can calculate, in advance, what arrangement it can make—unless you happen to have chosen some particularly simple example.

In general, we don't know how to do that, because once you have more than one type of atom you can make many different configurations, each of which will have its own energy. And of course what you're really asking me is, *Which one has the lowest energy?* There are so many of them. I'd have to be able to calculate them all, and then I'd have to figure out which ones have the lowest energy. But I'd find that the lowest energy ones are very different from one another, but with only very tiny differences in energy, so I'd have to be very precise in my theoretical calculation to tell which one is lower. I had better really know my atomic interactions very precisely.

But the differences in energy are much smaller than my uncertainty in those interactions, so I can't calculate it. This has always been a problem in our understanding of materials, and it's why material science has a heavy empirical component, a trial-and-error component.

For example, when Shechtman found what turned out to be an icosahedral material, it's not that he had the idea of *making* an icosahedral material: he was actually involved in a project—which by comparison is rather boring—of making many aluminum alloys.

Aluminum is a valuable material that has many different uses, so his project involved studying different alloys in the electron microscope to figure out which crystal pattern they correspond to, and then cataloguing them. He wasn't the materials maker, as it happens; he was the guy whose job it was to look in the electron microscope, measure the diffraction pattern, and record what the pattern was.

So imagine the situation: every few days someone is bringing him new materials and one day they bring him something which produces this strange pattern. He had the insight to stop and realize that he had chanced upon something really special.

Most crystallographers, upon seeing that pattern, would have discounted it. They would have just assumed that what was going on was the result of some multiply-twinning model that I just spoke about earlier.

As I said, this happens often enough in materials, so it wouldn't be unusual if what was going on resulted from multiple twinning, which would produce patterns that have "mock symmetries"—giving you the impression, at first glance, that you have a symmetry which is forbidden.

But if you narrowed your beam sufficiently, say, and instead of taking diffraction from the icosahedral nugget and instead focused on a smaller region within it, you'd suddenly see the apparently complex pattern simplified to that of a single crystal, again and again in different places.

So that's how you determine if it's multiple-twinned. Well, there's a lot of extra work involved to do that. So most crystallographers would have said, "*I'll just go on to the next sample and find one which isn't multiply-twinned.*"

In fact it may well have been the case that at the time Shechtman didn't actually know that much about multiple-twinning himself: my understanding is that it had to be pointed out to him afterwards that multiple-twinning was the likely culprit. Perhaps had he known more about it at the time, he might have just discounted it.

And in fact many of his colleagues *did* first discount it, saying, "*There are more tests that you have to do before you convince yourself you have something unique.*" So even though he first saw the material in 1982—I think it was—it took two more years before the case was strong enough that they felt they could publish the paper.

HB: Which might also mean that well before 1982, people had also stumbled upon the same sort of diffraction pattern and just automatically assumed that what they were dealing with was a case of multiple-twinning.

PS: Yes, we think that's likely the case. There are papers going back to the 1930s where the phase diagram of alloys is shown, and there's a phase in there that has a label which sits where the quasicrystal should be. It probably just wasn't recognized as being something unique.

HB: OK, so let's get back to your story now. There were three competing explanations...

PS: Yes, there were three competing explanations for a number of years. And at the outset the quasicrystal picture was disfavoured for a couple of reasons.

By this point people began to discover other similar diffraction patterns, but all of them shared the problem that they weren't giving pinpoint diffraction spots, and they weren't quite in the right positions.

They also had the common property that they were not stable. That is to say, if you cool those materials slowly, they would crystallize, but if you cool them sufficiently rapidly enough, they would form this particular structure. And that seemed to be consistent with the "icosahedral glass" picture.

Pauling got a lot of attention for the multiply-twinned picture—because he's Linus Pauling, and he deserved that reputation—but that one didn't convince many people because they could do this follow-up experiment I described of narrowing the beam and probing more carefully. And when they did that, they *didn't* find the characteristic signature of multiple-twinning.

And when people would look at the quasicrystal picture—when you put those units together that I showed you—they produce a pattern that looks something like this.

It's a very beautiful pattern, but also a very complex pattern. And the intuition was that nature is not going to make complex patterns. It would be much easier to make random, icosahedral clusters that just randomly aggregate. Now, this can't be *totally* random or you wouldn't get anything close to Shechtman's pattern—they have to align relative to one another.

That's not so simple either, but that was thought, at first, to be simpler and sufficient to describe the data, compared to the quasicrystal pattern.

Another issue had to do with going back to the Penrose tiles. If I give you a pile of Penrose tiles, you know—because of what Penrose showed—that if I follow his rules for how to join them edge to edge, I'm forced to make a quasicrystal pattern.

On the other hand, if I actually give you that job to do and you try to start building your pattern, you'll find that it's rather hard to make the pattern. And the reason is that, as you're going merrily along for maybe 10 or 20 units, you'll suddenly find that you're in a situation that can't be continued—you've "made a mistake".

You can correct it—you can move some and then make a different choice, and you'll get past that point—but in another 10 tiles, you'll find another mistake. And this will happen over and over again.

Now, if **you** have that problem, what about the *atoms*? How do **they** figure out how to make something quasicrystalline?

You could have gotten around your problem if you had a map, like what you have on the cover of a jigsaw puzzle box that shows you what to do. But that's cheating, because that's using long range information. Atoms don't walk around with little maps in their heads— well, they don't have heads, but if they had heads, they wouldn't be walking around with little maps in them.

In other words, it seemed like atoms needed long-range interactions in order to know, when one atom is attaching in one place, whether another should be attaching somewhere else. And in these types of materials, such as aluminum and manganese, we know that those atoms don't have such long-range interactions: they get shielded by the atoms nearby.

So that was a strong argument saying that, in principle, you could never do better than making something like a highly disordered pattern, much like an icosahedral glass. So in those first few years the stock in the quasicrystal picture went down, and the stock in the icosahedral glass picture went up.

Then two interesting things happened.

First of all, on the theoretical front, a group of us sat down with Penrose tiles in front of us and asked the question, *Okay, if we follow Penrose's rules, we understand that we can't go very far without making a mistake, but could we slightly improve those rules so that we can go a little bit further?*

It's a little bit like what I was saying about the icosahedra—and, if so, *how much further*, exactly? Maybe a little bit further, maybe much further, maybe a factor of 100 further. Perhaps we could go out far enough that people would be convinced that, at least on *that* sort of scale, we could do better?

I began this project, trying to do better than the Penrose rules, with Josh Socolar—the student I made that model with I spoke about earlier—George Onoda and Dave DiVincenzo who were at IBM. I remember we were sitting down at lunch one day to examine this question, and we made a bunch of tiles out of paper, and we began to make a tiling.

What would happen is that we would make a mistake, and then we'd look at the mistake and ask, *How could we have avoided that mistake just from* **local** *information—not from long-range information?* In other words, something happened along the way, could we have somehow avoided it? And then we'd see what it was and we'd write it down as a rule: *Whenever* **this** *local configuration comes up, don't do* **that**.

Then you start over again with the Penrose rules together with that new rule. Now you make it a little bit further, but then you run into another problem. And you repeat this process, adding another rule to your rule set.

Now each time you add a rule, it gets a bit more interesting: you're able to keep going longer and longer. Finally, after one more rule—I figure we must have had 10 or 12 by that point—we're not running into any more problems and we're able to fill the whole table with tiles.

So that was interesting. We didn't expect that. We thought there was going to be a steady, proportional increase to how far we could go as the rules increased, but suddenly we could finish the whole table.

Then we looked at the rule list. And the rules, as we had first written them, would have been something like, *If you have a skinny tile next to a fat tile at this angle, don't add it like that.*

But suddenly I realized that what we were looking at were rules that—unlike the Penrose rules, which constrained the way two tiles joined together along an edge—these new rules actually constrained the way tiles join around a vertex.

So it's still a local rule, but it's using not just your neighbours on either side of the tile, but a little bit more information. And if you viewed it that way, it was actually a very simple rule to describe. It was essentially: *If there's only one way to add a tile to a vertex, add it. If there's not, don't add it.*

We called these "forced" and "unforced"—either you had a vertex that was forced or unforced. And if you followed that rule, it seemed that you could avoid making mistakes.

Then the next thing was to put this on a computer and show that this was indeed the case. There's a little bit more to the story than that, but that's enough to capture the main thought of it. So it was this that got us over that psychological barrier that there was something about quasiperiodicity that required some sort of long-range conspiracy.

Even though you might have had that impression based on Penrose's tiles and based on your experience, in fact it's not true. And that's very interesting and surprising, and still not completely understood. I still work on aspects of that.

Now, around the same time, independently, a group in Japan discovered another quasicrystal, but unlike the cases of Shechtman's and the other ones up to that point, this one could be grown in large nuggets, beautifully faceted, with diffraction spots that were pinpoint: when you measured them—they were dead on to what the quasicrystal picture would have predicted.

And these materials had the advantage—which is why they could be created—that they could be grown very, very slowly, the way we grow very highly perfect crystals. They didn't crystallize. We don't know for sure if they're stable to all temperatures, but as far as we can measure them in a laboratory, they seem to be stable.

That basically changed the mindset of people overnight: the icosahedral glass community immediately conceded that they couldn't account for something with such a perfect pinpoint diffraction pattern with their model.

Pauling was a little bit more stubborn, but I think he realized that if he was ever going to make a pattern that was as perfect as that, he suddenly needed a very complex model with no good reason behind it. Whereas before he had used repeating units of hundreds of atoms, he now needed to go to tens or hundreds of thousands of atoms. So that essentially established that quasicrystals can be synthesized in a laboratory.

HB: These guys in Japan—how did that come about? Was it, again, serendipity? Or were they deliberately trying to create something like this?

PS: Yes, they were doing it deliberately. The question, *Could you make something more perfect?* was definitely out there. While the theoretical prejudice was developing regarding why you couldn't make something perfect because of this long-range conspiracy, they fortunately just charged on and started looking for other quasicrystals.

Like most work in material science, when you're looking for something new, it's mostly hunt-and-peck, trial-and-error, serendipity, with a little bit of intuition.

And the intuition they were using was that the first material that had been found had aluminum and manganese—so the first things that people began focusing on were aluminum plus other materials with atomic properties similar to manganese in hopes of making something better.

At first they just replaced the manganese with another single type of atom, but this example was aluminum, iron and copper: they had used two different elements to replace the manganese. Even that's not enough, it turns out—you have to play with different combinations, run through many different combinations of various ratios of them. And then you have to look through the material, which may or may not be homogeneous, to find if there's something interesting there. So it's a long, painstaking process.

But that's what material scientists are used to doing, and that's what makes it a very challenging field. But eventually they did find a new material which was very, very interesting.

HB: So when this happened, it must have been a big deal.

PS: Yes, absolutely. I would say that they found the first, bona fide quasicrystal. Shechtman had found something icosahedral, but we couldn't quite be sure what it was. But this was the first material that took us past the point where there was no more ambiguity about whether or not quasicrystals exist.

Questions for Discussion:

1. Why do you think that the "energetically preferred" structure necessarily implies one with the lowest amount of energy?

2. What does Paul mean, exactly, when he says that access to a map, like you have on the cover of a jigsaw puzzle, provides "long-range information"?

3. Does "scientific intuition" exist? If so, what is it, exactly and to what extent do you think it is different from any other form of "intuition"?

4. To what extent does this chapter provide an instance where "falsifiability" plays a clear and distinct role in successfully distinguishing between competing scientific theories?

V. Looking to Nature

Developing a separation algorithm

PS: So it was clear to almost everyone at this point that quasicrystals exist, and the question now was, **Why** *do they exist?*

And here began a debate that some people still engage in today. Is it because they're inherently delicate forms of matter that, if you form just the right conditions in a laboratory, you can manage to make them, freezing the material just before it crystallizes? Or are at least some of them—just like crystals—actually *energetically-favoured* phases of matter? In most cases, probably, crystals are favoured, but maybe, for some combinations at least, quasicrystals are favoured.

If you have something in mind like the models that I was showing you before that have those Lego-like joinings that would mimic the atomic interactions, perhaps then the quasicrystal is favoured and you can't make the crystal.

Are materials like that? Is it simple energetics—just like for crystals—that forces them to be what they are? Or are these particular examples simply very delicate forms of matter? That's important, both for fundamental physics and also for application: if you're going to use the materials for something.

A key thing for us was the fact that we could find these so-called matching rules, or forcing rules—these "growth rules," for growing it perfectly. The fact that you could do that meant to me that, somehow, nature would take advantage of that, at least for some materials.

So while some might be delicate, there was no reason I could think of why they should *all* be delicate. And if they weren't all delicate, then there's no reason why you necessarily had to make it in a laboratory—maybe nature made them also. Nature also made crystals that aren't always stable—like diamonds, which are stable under the earth, but not so stable when they get to the surface of the earth—so maybe nature made quasicrystals too.

From fairly early on, even when we were just beginning to develop the idea, I already had the idea to look in nature, rather than to try to manipulate colloids or other laboratory techniques. Maybe nature had made them and we just hadn't noticed them.

So whenever I had a chance, I used to go to museums and just look in mineral cases, thinking, *Maybe someone's mislabelled something, maybe they're icosahedral things.*

I went to The American Natural History Museum in New York and the Smithsonian in Washington, both of which have two very extensive collections. The problem there, for me, was that you actually do find icosahedral things, but they all turn out to be these multiply-twinned materials.

The most famous example besides gold is fool's gold—pyrite. Fool's gold will often form beautiful cubes, but sometimes you'll also see it form these nuggets that look like they're made of pentagons on the surface. So you think, *Oh, this violates the laws of crystallography. It must be a quasicrystal.*

But then you look more closely and you realize, "*Actually, those pentagons are a little distorted. They're not quite right.*"

This is actually something that is quite well known in the study of pyrites: they form this second form of cubic lattice where the facets are pentagonal—distorted pentagons—but it's still a cubic structure, which you can see if you look closely at its diffraction pattern.

It turns out there are a bunch of materials like that: I learned all about what are called pyritohedra and other things that would superficially mimic a quasicrystal, but I didn't find anything that was promising that way, so I put that aside for a while in my mind.

HB: These guys at the museums just let you come in?

PS: Well, they're public museums, so you can look at the collections —I didn't go to the back rooms or anything.

HB: Oh, you were just looking in the public rooms?

PS: Sure, there are lots of materials there. It wasn't a very intelligent way of searching. It's just something you do. You do the best you can.

Occasionally I would run into somebody who was working in one of those museums and I'd ask them if they had seen anything that reminded them of what I was looking for. It was just one of those things I would occasionally come back to.

But I discovered something a few years later that made me think there was a way of searching systematically for them. It turns out that there is an international Catalogue—a computer database—of what are called powder diffraction patterns.

Powder diffraction patterns are not like the ones I've been describing to you up to this point where you take a single crystal and you shine electrons or X-rays through them and you get that beautiful lattice of pinpoints.

In order to do *that*, you need a very perfect, large, single grain of material, and often for materials, whether they're synthetic or mineral, you're not so lucky and instead you have many tiny little grains, none of which is big enough to give you a diffraction pattern.

You can take the diffraction pattern of all these grains—it's called a powder pattern—where some of the crystallites are rotated one way and some another way, and so forth. So the pattern you get isn't that beautiful lattice of pinpoints like I was discussing earlier, but instead imagine taking that lattice of points and rotating it around its centre, so every point now traces out a circle, and you get a lattice of circles.

You can't tell the symmetry of those patterns just by eye, but there's still information there: the spacing between those circles is actually information. For example, if it's a crystal, the spacing between the radii of those circles has to obey certain relations. If it's a quasicrystal, you're going to get other relations, because it has that quasiperiodicity—those irrational ratios—in there. So I had the idea that, somehow, by looking at the powder diffraction patterns, one might be able to identify some quasicrystal candidates. And since there were something like 9,000 mineral patterns in there, I would be able to look at 9,000 at once.

I had that idea in my mind, but I was working on other things. Then, shortly after I came to Princeton in 1998, I was asked to give a colloquium and decided to give one on quasicrystals and what we knew at the time. Sitting in the audience was a geoscientist by the name of Ken Deffeyes, who was then on the Princeton faculty. He came up to me afterwards and he said, *"Great talk. But I'm curious, has anyone ever seen a natural quasicrystal before?"*

I said, *"No, but I think I know how to look for one,"* and that got him really interested. I described a little bit about my idea to him, and he said, *"Oh, I have just the person for you to work with. There's a bright undergraduate student by the name of Peter Lu. He's a physics major, but he's also a mineral collector who won all kinds of contests for his understanding of minerals when he was in high school. And he knows how to use an electron diffraction microscope."*

So a few days later, I met with him and he decided to work with me on his senior thesis project at Princeton. The first goal was to find the mathematical algorithm of these powder diffraction patterns that would help pick out the likely quasicrystals—in this case icosahedral quasicrystals—from other things.

It took a while to get that method to work, and we could test it by using it on examples of real quasicrystals that at this point were already known synthetic materials.

Then there were other samples that, in our separation algorithm, were close to that of the quasicrystal; and those would be our candidates.

For each one, we had to find a sample of it, bring it to Princeton, slice it, dice it so it's thin enough for the electron microscope, and then rigorously test it to find out whether it's a quasicrystal or not.

You can imagine that's a very tedious process—even just finding suitable candidates. It was never the case, in any of the examples, that I could just go to the Museum of Natural History and get a sample of it. It was always only in just *one* location of the world, and it was usually some place where only *one* person had it in some back room somewhere. We had to somehow find it, bring it back here, slice it, dice it and put it in the electron microscope to invariably discover...

No, it's just a crystal. Then we'd try again with something else. And again. There were all sorts of funny adventures along the way.

At the end of about a year and a half, we hadn't come up with any examples—they were all false positives. The problem wasn't so much our algorithm, but it turned out that the International Mineralogical Catalogue isn't precise enough.

The Catalogue collects data from various groups, but the precision of the instruments has improved over time. And when we look with more precise instruments, we see the circles aren't quite the radii that were reported. They were just enough off that, while in the Catalogue it seemed to be a good quasicrystal candidate, in reality, it wasn't. You have to live with false positives if you're going to play this game.

So we failed, but we wrote a paper at that time explaining our methodology and our failures in detail. But we still had things on our list that we hadn't yet been able to find. By now Peter had graduated and was going to Harvard to work on a PhD. Ken was emeritus and had left to go to San Diego. So it was basically left to just Nan Yao, who does the electron diffraction work in our imaging lab, and me to continue the project.

At the end of the paper we specifically asked for outside help to best explore additional candidates, but nobody responded.

Then, six years later, I get an email from someone I'd never heard of: a fellow by the name of Luca Bindi. He explained that he's the curator of the Mineralogical Museum in Florence and he'd be happy to join the search.

So I found another crazy person.

Questions for Discussion:

1. What do you think Paul means, exactly, by "delicate forms of matter"? Might there be some technique to render something "less delicate"?

2. Can you explain in your own words why powder diffraction patterns necessarily involve circles?

3. What, if anything, does Luca Bindi's response indicate about the global nature of the scientific enterprise? How does the fact that it took him 6 years to respond to Paul's explicit quest for further assistance illustrate the timelines of interaction that science operates on?

VI. New Year's Delight

Persistence pays off

PS: It turned out to be quite fortunate that, of all the people in all the world, Luca was the person who joined the search, partly because—as it will become clear in a moment—there was something interesting in his own museum, but equally importantly—probably more importantly, in fact—was that Luca turned out to be as fanatical about this search as I was.

There were many instances over the following years where we got stuck—just dead in the water—and it took some combination of stubbornness and luck to get past it. More often than not, it was Luca who figured out the way to get us past that barrier. He's also marvellously skilled in the laboratory in terms of picking out microscopic materials that you can't see, putting them on the head of a pin and moving them from one place to another. He's just a really amazing character.

We began a relationship that has only grown since then: we Skype with each other almost every day, because there's always something going on in this subject.

But in the beginning, that wasn't so clear. For the first year or so, it was just like it was before with Peter: Luca would find a material that was in the Catalogue that he happened to have in his mineralogical museum. Then he would slice and dice it, do a crude measure, and it would turn out to be a failure—over and over again.

Then, about a year into it, he said he wanted to try something that wasn't in the Catalogue at all. The Catalogue only collects certain minerals, and he had some minerals that were not in the Catalogue. One of them looked particularly interesting because it had metallic aluminum in it, which was known to occur in some of the quasicrystals that had been synthesized in the lab. So we started to look for something of that sort.

We found some material called khatyrkite. It came in a small box that described it as coming from the Koryak mountains in far Eastern Russia. And in the box was this tiny grain of material, just a few millimetres big. That ended up being a very important few millimetres.

What did Luca do? Well, just like for every other sample, he sliced the material, so that we could later probe its local chemistry with the electron microscope. It was a rock, not a nugget, so it had many different minerals in it, and only some of them were really khatyrkite.

Khatyrkite turns out to be a crystal phase of copper and aluminum—$CuAl2$. Other regions showed a different crystal phase known as cupalite. He found two that didn't correspond to anything in the Catalogue among the known minerals. So those would be the things to go after.

One of them turned out to be a crystal phase that we eventually understood, while the other one looked more promising. So he tried to isolate that. In this case "isolate" means "pick out those tiny, little grains."

By the time you've done all this, the sample is gone except for these two little grains. So when it finally came to me in a box, there were just two glass needles, at the ends of which were two little specks of stuff.

That wasn't exactly what I had in mind: I was looking for a quasi-crystal the size of my hand. This sort of thought frequently occurred to me while looking at different samples: after slicing and dicing them I would think, *If there turns out to be quasicrystals inside this, I'm going to be disappointed, because I wanted something much bigger. I can barely even see what I'm looking at here.*

Anyway, I brought the sample over to Nan Yao in our PRISM lab at Princeton. We tried to do some measurements while they were still on the glass needles, but those didn't work out so well, so we had to try to remove the little specks from the glass needles.

The way Luca had prepared the samples was that he had punched out these little grains, put some glue on the needle and then basically dipped the needle in the stuff. So we figured we would just unglue it. Nan had the idea of putting some acetone on the end of it. We put one, tiny drop of acetone on the end and poof!—the whole thing disappeared. And I thought, *Oh my God!*

Fortunately, there was a little crucible below which the sample fell into, so we actually lost nothing. If, for some reason, we hadn't had the crucible there, that probably would have been the end of that project. But very fortunately it was.

Now, the next problem was that we had to put it in the electron microscope, and the problem with that is that the electron microscope was booked up for several months.

HB: Several *months*?

PS: Yes, our electron microscope is a shared facility, and many people sign up for it. So we had to go in on an off-hour, and that off-hour turned out to be at 5 am on New Year's Day, 2009.

I'm sure both our families thought we were a little bit nuts, but we snuck out of our respective houses, converged in the lab, and by the time I had gotten there Nan had managed to get the speck in the electron microscope—the speck itself was made of the powder of even tinier grains, so he managed to find one that was thin enough that the electrons could penetrate part of it.

And very shortly afterwards a diffraction pattern appeared that was a beautiful quasicrystal diffraction pattern: pinpoint-like, beautifully aligned, much more perfect than Shechtman's sample.

So we just knew immediately that we had found a quasicrystal in this rock.

Questions for Discussion:

1. How common do you think it is for a theoretical physicist like Paul to be so directly involved in handling and working with materials?

2. Are you surprised by the amount of passion and dedication exhibited by everyone involved in this project? Do we do a good enough job describing such passion to young people to illustrate to them the thrill of scientific discovery?

VII. Confronting the Impossible

Encountering rock-hard scepticism

HB: You strike me as having been incredibly tenacious throughout this whole experience.

PS: You had to be. By this point there had been tens of adventures that you had to be absolutely crazy to undergo. Bad things would happen, and you'd have to compensate for the bad things, and so on. The whole nature of the project was that you had to be nutty and fanatic to pursue it, because on many levels it required a kind of fanaticism in order to move forward. Otherwise, you shouldn't even start because you're looking for a needle in a haystack—in fact, worse still: maybe the needle isn't even *in* the haystack.

HB: But now that you've seen this thing, you must have been thinking that there must be bigger quasicrystals somewhere, right?

PS: We thought we should be finding more, even in the grains that we have. The first thing we wanted to do was figure out what the symmetry of this material was. We had found a quasicrystal diffraction pattern that was similar to Shechtman's but didn't necessarily have to have the same symmetries. To check if it has the symmetry, you have to rotate the material, find other diffraction patterns, and figure out what the other symmetries are.

But, sure enough, it *did* have the same symmetry as Shechtman's icosahedral pattern: it was icosahedral. The next thing we did was measure its chemical composition, which turned out to be something like 63% aluminum, 24% copper, and 13% iron.

And when I saw those numbers, I thought, *Whoa, I know **that** combination. That's the same combination that the Japanese group in 1987 had synthetically made in the laboratory—it's the same material.*

In fact, I had some of that synthetic material in my office because I had worked on that material when I was at Penn and had kept some of it; and we were even able to compare one with the other to really check the composition carefully. They really were the same.

The only thing was that An-Pang Tsai, the group leader in Japan, went trough a tremendous amount of effort to synthesize that, while *our* sample was a grain in the middle of a complex rock with all this other stuff. You would never think to make a highly perfect quasicrystal in the middle of something like that.

The project could have just stopped there with an announcement paper. I called Ken Deffeyes to tell him the news—he was now emeritus, living in San Diego. I said to him, "*I have a question: How in the world did nature figure this out when we have to work **really hard** to make it? What is this telling us about quasicrystals? Is it something about their stability? What is it telling us about some particular geophysical process?*"

He said, "*Well, there's a fellow you should go see,*" and he recommended a famous petrologist here at Princeton called Lincoln Hollister—a petrologist is someone who specializes in the way rocks physically form. So I contacted him by email, made an appointment with him to meet in his office—which is just a few blocks up the street from where we're sitting now—and I told him the whole story I just told you now. And he gave me this look which I later learned meant *Watch out!*—but at the time I didn't know what it meant.

And he said to me, "*Well, what you have there, Paul, is **impossible**.*"

And I replied, "*No, I don't think it's impossible, because we know we can synthesize this in the laboratory, and we know quasicrystals are possible.*" And I started to tell him all about what we knew about quasicrystals.

But he interrupted me and said, "*Well, I don't know about the quasicrystal part. That's new to me. But what you've told me is that the material you have there is a mixture of metallic aluminum, metallic copper, and metallic iron. Metallic aluminum is known to have a voracious affinity for oxygen. There's lots of aluminum in the earth, but it's all attached to oxygen, and we actually go to some effort to separate the two—that's what aluminum foundries are for—and it takes a special process to do that, a lot of energy. But in nature, we don't find aluminum metal. It's impossible."*

Well, when that word "impossible" comes to me, there's a question I always ask, which is, *"When you say it's impossible, do you mean it's **probably** impossible, like 1+1=3? Is it **that** kind of impossible? Or do you just mean that it's very, very unlikely? In other words, does it simply violate some common assumptions, or are you telling me that it's really, physically impossible?"*

He thought about it for a few moments, and then said, *"Well, if I were forced to come up with an idea for such a material that was natural, I'd need to get conditions which are highly reducing to strip the oxygen from aluminum. And the only place you're going to find that is some place deep under the earth, near the boundary between the core and the mantle. And then, you have to figure out—if you manage to do all that—how you are going to get it to the surface of the earth."*

And he added, *"Some people have speculated that there are these things called superplumes, which perhaps had once extended down to the core-mantle boundary. That could possibly give your sample a ride up to the surface."*

OK, I thought to myself, *He meant the **second** kind of impossible— it's a very unlikely story, but if it's true, it's really interesting.*

In fact, it symbolized the first proof of such a theory, if that were the case. So it was really interesting, albeit unlikely.

For some time, I'd been thinking about an alternative possibility that I asked him about: *"Well, how about meteorites?"* I was very naive about them, but I thought, *Well, there's no oxygen in space*—that's not true, as it happens; in fact, there's lots of oxygen in space—*could it have been made under those conditions?*

And he said to me, *"I don't really know that much about meteorites, but I know someone who does."* So a week later we found ourselves on a train to the Smithsonian in Washington to visit the head of the

Division of Meteorites, Glenn MacPherson. We get to Constitution Avenue, we're about to enter the door, and Glenn's already at the door. He's very excited and wants to tell me, before I even hit the door, that what we have is impossible.

Both Lincoln and Glenn are these pretty tough guys—this field is full of very tough, sceptical people—and as sceptical as Lincoln is, Glenn is at the next level. He began to tell me, during the long walk to his office, all the various reasons—not just the aluminum part—why what we have can't possibly be from a meteorite. He thought that it was certainly going to be a piece of slag, an anthropogenic byproduct of some sort of industrial process.

Several hours later we found ourselves back on the train. I'm sure that neither Glenn nor Lincoln thought they'd ever hear from us again. But he didn't account for the fact that Luca and I were both pretty fanatic by this point. Even if it were something anthropogenic, I still wanted to understand how.

I felt like it was a win-win situation. Of course the outcome wouldn't have been nearly as nice if it didn't turn out to be natural, but either way we really wanted to try to understand how this thing had actually formed.

Questions for Discussion:

1. Do you find Paul's distinction between "two types of impossible" a helpful one? How often do you think professional scientists confuse one type with the other?

2. To what extent do you think that the sceptical attitudes of Lincoln and Glenn enable them to become better scientists? To what extent do you think their reactions also might have helped Paul and Luca develop and refine their ideas?

VIII. Tracking Khatyrkite

Smoke, mirrors, and the holotype sample

PS: So began another adventure, which basically took the next year and a half, to try to find out how the sample had reached Florence in the first place, as well as further test it to rigorously compare it to industrial slag.

According to the box, it didn't come from an aluminum foundry: it came from some remote place in the middle of nowhere, where there are no foundries. Well, was that actually true or not?

In the meantime, the laboratory still had a few specks of the sample left, and we wanted to know if, by studying those specks and comparing it to slag, we could distinguish which was which.

It's impossible to describe how the next year and a half worked out, because literally every day, there was news on one front, or the other front, or both fronts—sometimes fluctuating good news, sometimes bad news. Sometimes even within the same day there was something happening.

In the meantime, Luca and I had other things going on in our lives—we were working on this on the side—which is why we began these daily debriefings by Skype.

It gradually became a little bit competitive, in fact, regarding who would have the most interesting news of the day. Some of it was about literature searches, searching for people, and things like that.

HB: So it was just the two of you?

PS: Basically just the two of us, yes. Every now and then—although Lincoln and Glenn were sure they weren't going to hear from us—I would send them information to see if we were getting closer to switching their needle from negative to positive, or giving us more information. So we kept them in the loop. And Nan Yao was also still involved in the lab work.

I'll focus now on the story of where the rock came from. The box says it's khatyrkite—and, of course, we've verified that it has khatyrkite in it—and the only other thing it says is that it comes from the Koryak Mountains, which are in far Eastern Russia.

So how would you go about checking to determine if the rock *did* indeed come from there? Well, he first thing you'd probably do is ask, *Is there any record of how the rock got into the museum?*

It turns out there was one sheet of paper, and that one sheet of paper explained that the rock was sold in 1990 with 10,000 other samples by some particular collector who lived in Amsterdam.

OK, so you've got a name and are keen to follow that lead to try to find the collector. Fortunately, thanks to the Internet, you can now search directly through Amsterdam telephone books—and even search the streets of Amsterdam—to find mineral collectors.

Well, we did that for several weeks but we found no such collector. I got some sort of hint that someone with that name was alive in the 1990s, but nothing after that point. Once again we had come to a dead end.

Then we tried the next idea, which was to try other museums. After all, this sample happened to be in the Florence museum, so perhaps it would be in other museums too?

We tried to contact different museums, collectors, people who sell minerals—all in an effort to track down other khatyrkite samples. And we got some hits. We got about three in the West and one in St. Petersburg, Russia. The ones in the West were kind enough to send us their samples. But when we examined them in the laboratory, we found nothing.

And when I say "nothing," I mean that not only did they not contain quasicrystals, they didn't actually contain khatyrkite either—they were fakes. It turns out that there's actually quite a bit of faking in the mineral market. Minerals are collected by collectors and are relatively inexpensive to buy, but relatively expensive and time-consuming to test.

The kinds of things we were doing—slicing and dicing them, and isolating the sample—involved months of work for each sample. People don't tend to do that. They tend to buy the sample, and then if someone is stupid enough to write them and volunteer to test their sample, they'll send it to you. And they'll take the risk that they'll either win or lose. And unfortunately, these guys lost.

Even today, thanks to the story I'm telling you, if you go to one of these big shows in Tucson or Munich where they sell minerals, you'll see khatyrkite on sale, but I suggest not buying it because all the samples we've tested are fake.

HB: So this stuff is very rare, presumably.

PS: Yes, it turned out it was very, very rare, which we didn't know. That was very costly in the sense that Luca had sliced and diced and pulverized it, and lost all the context around our sample in order to go for his one bit, thinking, *Okay, this won't be so hard to replace.* Of course, this infuriated a traditional petrologist like Lincoln.

I should say that in this field there are the mineralogists, who care about one mineral in particular, and then there are the petrologists who want to know the context of that mineral, relative to others. For them, it's important to keep the things intact, and for us it was important because we wanted to understand how our sample formed. So this cost us a lot of time.

On the other hand, the sample in Russia—we knew *that* one was real. And we knew it was real because it was the holotype sample for khatyrkite. If you're a mineralogist and you think you've discovered a new mineral, you write up a little report, present the data to an international commission and there are representatives from 30 countries who vote on whether or not your data are good. They might push back and ask for more information, and eventually they might accept it.

And then, you're ready to declare that you have a new mineral, but for that, you still have to do two more things: you have to publish a paper about your mineral and you have to put a sample of it in a museum. And the one in St. Petersburg was that museum sample.

So we knew that one was the real deal, and we even had the paper that described some of its properties, and although it didn't mention anything about quasicrystals, it told us something about it.

HB: So you have the bit in St Petersburg—which you know, tautologically, is the real deal—and then you had your sample that happened to be in Luca's museum.

PS: Right. But we don't know if they're related. We just know that they both have khatyrkite in them. One could have come from the South Pole and one could have come from the North Pole. You just don't know.

Now, it's true that the sample in St. Petersburg was listed as coming from the Koryaks—they had described where they had found it. The Koryaks are in the Kamchatka Peninsula, at the far Eastern edge of Russia, just the opposite side of the Bering Strait from Alaska.

The part the sample came from isn't the southern part, which people are more familiar with—it sticks out into the Sea of Okhotsk and has volcanoes, which tourists sometimes visit—but is instead from the northern part of the peninsula, which is desolate and has about a hundredth the population of the Western Sahara.

Due to its proximity to Alaska, it's the closest boundary between the US and Russia. It almost started World War III during the Cuban Missile Crisis, and now it's more of a mining area. Anyway, that's where our sample came from.

We were thinking that if we could prove any one sample was natural then that would be enough to get Lincoln and Glenn to rethink their positions. Of course, they don't trust those guys any more than us—maybe less than us, in fact—so we had to find out if the paper that was describing this other sample was correct or not.

Ideally, we'd check the material directly, but you can't check that material because it's the holotype material. So instead we tried to track down the authors of the paper.

It turned out the lead author on this was a fellow by the name of Leonid Razin. And we eventually discovered, through searching on the Internet, that he was head of the Platinum Institute during Soviet times, which makes him a very suspicious character regarding what his political connections would have been. And it turns out that he *did* have those kinds of political connections.

We heard rather bad stories regarding the way he would treat his competitors using his connections—questions about his character that would make you very worried about whether this sample is real or not. At any rate, I eventually found out that he had emigrated to Israel.

So now I start looking on the Internet through telephone books in Israel and eventually I find some guy named L. Razin. I call him up and no one speaks English. So I call him up with someone who speaks Hebrew, and no one speaks Hebrew at this place either. The third time I call with someone who speaks Russian, which does the trick.

I ask him if he's Leonid Razin, and he tells me that he is and confirms that he is the guy who, in 1985, wrote the paper describing khatyrkite and found the original sample.

It was all going well. But then I asked, *"Can you describe the conditions under which you found it?"* because the naturalness very much depends on where somebody finds it and under what conditions.

It turns out he didn't really remember that much, and what little he *could* remember was exactly the stuff that was in the paper, which was not very much at all. That was worrisome. Then I asked, *"Do you have your geological notebook?"* because that would have all those kinds of details in it. And he replied, *"I'm not sure. It might be in Moscow."*

At this point I began to think, *Hmm, this is bad*, because most field geologists **always** take their notebooks with them wherever they go. So I asked him, *"Well, do you have any more material?"* and he replied, *"Maybe in Moscow."*

At this point during the conversation, I was looking at the cost of airline travel from Tel Aviv to Moscow, and it wasn't so bad. So I asked him, *"If I pay your way, would you be willing to go back to look for the notebook and more materials?"*

The discussion got kind of complicated at this point, and it took us a few days of back and forth with various people getting involved to figure out that what he wanted was a significant monetary reward if he were to go back and do that. I tried to explain to him that this was a project that was operating with zero federal funding—everyone was working for free on this project and we had no such resources.

But that didn't convince him. He just got angry when I said that. He was a tough character to deal with. We had to make a decision as to whether or not we should even try the whole approach.

Some people told me, *"You're at a dead end unless you decide to pay him."* I thought about whether or not I should try to come up with some resources to pay him, and I figured maybe he'd lower his price, but I got worried because he wasn't able to answer any of my questions and I was afraid he would come back with a notebook and no materials. And if he *did* come back with a notebook, how would I know when that notebook was written?

After a few days, I decided that I simply couldn't pursue it. So that was the end of that lead. We were out of leads. That's an example of what I was saying before: there are various points in the story where we just thought we were done.

HB: OK, fine—you've convinced me that this was all enormously difficult and that you're an incredibly determined fellow, but don't leave me hanging here.

PS: Oh, you were expecting something interesting to happen? OK, I guess I'll continue then.

One night Luca is having dinner with his sister and a friend, telling the story that I just told you, and the friend asks, *"What is the name of the actual collector of your sample?"* Luca tells him, and he says, *"Oh, I live in Amsterdam, and unfortunately that's a very common name, like John Smith. As it happens there's an older woman who lives down the street—I help her with groceries—and she also happens to have that name. I'll go ask if she knows anything, but it's not likely."*

Twenty-four hours later, we get an email and—guess what?—this woman turns out to be the widow of this collector. So the good news is that we found the collector of our sample. The bad news is that the collector himself is no longer alive.

So off Luca goes to Amsterdam to meet this woman. She won't meet with him, but she'll meet with his Amsterdam friend—he has to do this indirectly, because she's very nervous. Anyway, she explains that her husband was indeed the collector and he used to collect minerals and seashells. At one point, back in 1990, he sold his minerals because he just wanted to collect the seashells. Other than that, she really knows nothing about it.

It doesn't sound very promising, but after enough questions, she admits that her husband used to keep a secret diary. And although the minerals were sold to Florence, she kept the secret diary. She lets Luca's friend look at the secret diary, and sure enough, there's an entry in there for khatyrkite involving a trip the collector took to Romania in 1987 where he met someone simply described as "Tim the Romanian."

Now remember, smuggling minerals was strictly illegal in Soviet times, so "Tim" was clearly a mineral smuggler, and that's how the collector got his minerals. When I got that news, I thought, *Wonderful—now we're past the collector, and surely there can't be anything easier than to find Tim the Romanian mineral smuggler.*

So back to searching the Internet again, and the same thing happens: nothing. No one has ever heard of anyone named Tim, anywhere, ever, using that as a pseudonym or anything, for mineral smuggling. No one has ever heard of mineral smuggling in Romania, which certainly took place, but nobody officially knew anything about it. So we were dead again. Should I go on?

HB: Come on.

PS: Well, what can we do? We send Luca back to Amsterdam to ask if the wife has ever heard of Tim the Romanian. Of course she hasn't, because she had nothing to do with his collection, so she knows nothing about it. But she finally confesses that her husband used to keep a "secret secret" diary.

And it turns out that in his secret secret diary, he kept details of his illegal dealings. And from this we discover that Tim the Romanian was actually getting his stuff from a certain laboratory owned by a person named Rudashevsky. Well, we knew *that* name—that was the same laboratory where the analysis was done for the St. Petersburg sample, the holotype sample. And it goes on to explain that Rudashevsky was getting his materials from Leonid Razin—our guy in Israel.

So in fact our guy Razin—now in Israel—who was well known for getting competitors in trouble by accusing them of smuggling minerals out of Russia, was *himself* smuggling minerals out of Russia. But more importantly, we now know our material wasn't just "sort of like" the one in St. Petersburg; it is *part of* the one in St. Petersburg.

So if we could figure out where *that* one came from, we can then, with confidence, be sure where ours comes from. It looks good now, because at least according to the paper it really *did* come from the Koryaks. So it wasn't just on the box, it wasn't just an accident: it really turned out to be true.

But at this point we no longer believed that Razin had anything to do with this discovery—somebody else must have brought it to Razin. So who did *that?*

So we spent a long time asking various people who knew him trying to figure out who he had used—again, there were lots and lots of dead ends of various sorts. I tried to approach Razin again through Rudashevsky—whose son, in particular, was very interested in helping us.

He called Razin and tried to explain to him that this was a scientific investigation, but Razin basically blew up at him and accused him and his father of being smugglers, and told him that he was going to use his former KGB connections to get them in trouble, essentially threatening them with their lives.

And although they didn't take it seriously, *I* took it seriously. So I didn't want to go to Israel to try to press things further there. And there was no point anyway, because I didn't think he could help us much more anyway at this point.

HB: Right. So then what did you do?

PS: Well, at the beginning of the original article, there was a description, in very vague terms, of the surrounding mountains where the sample was found. And there's mention of some guy washing clay from this stream who is even named—Valery Kryachko—although he wasn't listed as an author on the paper.

It was a weird business. It wasn't exactly clear what was going on: there weren't enough verbs and nouns in this description to even be sure it was talking about how they got the sample—it wasn't exactly clear *what* it was about. But among the many leads we found, we tried to trace the guy who was explicitly mentioned.

Meanwhile, we'd been told by members of the Russian Academy of Sciences that this was going to be a fictitious character, because if you were the head of the Institute of Platinum and you were explaining where you found khatyrkite, which is effectively worthless, you're not going to give away where you *were*, in fact, looking for platinum.

This was a common technique, we were told. We had also been told other things—someone else said that he was a real person but he was dead. So we had given up on this lead a number of times.

But then, one day, while pursuing one of our other many leads, we ran into a 1995 paper on the Internet, on which he was listed as an author. It didn't tell us how to find him, but we *were* able to find one of the co-authors.

I managed to get one of the co-authors on the phone in Moscow, whose name is Vadim Distler, and I asked him, *"What about this guy Valery Kryachko, your co-author on that 1995 paper? What can you tell me about him?"*

And he told me that he was indeed a real person—which was obviously good news. In fact, it turned out that he'd been Distler's PhD student—not at the time of the khatyrkite discovery, as it happened, but after he had worked for Razin for some time—as it happened,

Distler had heard horror stories about Razin from Kryachko's experience of working with him.

And then he told me that he was still alive, which was wonderful. And *then* he told me that he was visiting him the following week and asked if I wanted to talk to him myself. So this guy suddenly went from being an imaginary character to being someone I could actually interact with. That was extremely exciting.

And our interaction, like most of my international interactions, was done by email, using Google translate back and forth.

Valery was wonderful straight away. He wrote me a detailed story about how, in 1979, he was a Master's student and was sent by Razin out to this remote place in the Koryaks where there was this tiny stream in which gold had been found. Now, gold is often associated with platinum, so his job was to look for the platinum.

He and one other guy had spent a week there looking for platinum by digging stuff out of the stream and panning it, but had found nothing. He didn't want to go back to Razin empty-handed because he knew that was going to be big trouble.

But he had found these shiny things, which he knew wasn't platinum, but at least that way he could give *something* to Razin. So he gave them to Razin, and that's the last he had heard of this up until I contacted him. He didn't know that Razin had taken those samples back to St. Petersburg, analyzed them, found new minerals, published the new minerals, or that he had smuggled anything out.

The only thing he had heard about the story—because by now, our first paper had come out announcing that there might be natural quasicrystals—was that there might be a Russian connection to that, but he naturally didn't realize that *he* was connected to the story. But of course he was.

He was really excited about that. He's a really wonderful person generally, and he volunteered to do anything he could to help. For us, the most important thing was that we finally knew exactly where, when, and how our sample was found. It was **not** found in an aluminum foundry. It was a real, natural, something.

Meanwhile, the story I *didn't* tell you was everything that happened in the laboratory. But let's just imagine that, after lots and lots of work on these little specks of material, what we found in the laboratory was that our sample had been formed under very high pressures, pressures probably 100,000 times atmospheric pressure—again, not something you'd find as a byproduct of any kind of human activity.

At one point, we had roughly a dozen ideas, but we were now back to our original two ideas that the only ones that were compatible with high pressure were either deep under the earth or in space.

HB: The superplume theory or the meteorite.

PS. That's right. And at the time it was anybody's guess as to which one it was. But it turns out that there's a nice way of distinguishing the two, which is by measuring the oxygen isotopes in the material. It's an expensive process, so you don't go right ahead and do it first thing, but there are a few places in the world where you can do it, including Caltech.

By this time, I had been travelling back and forth to Caltech, trying to get people to see if they had any ideas, and one of them put me in contact with John Eiler and Yunbin Guan there, who do these kinds of measurements. And now, finally, there was good reason to do those measurements to distinguish the two possibilities.

It took about a summer's worth of work to do it, but by the end of the summer the data was clear and unambiguous: it corresponded perfectly and beautifully with meteoritic origin. Not only can you tell it's a meteorite, you can tell *which kind* of meteorite. It turned out that it wasn't just any kind of meteorite, it was a special class of meteorites known as CV3—carbonaceous chondrites.

For meteorite experts, these are the most interesting and famous ones because they're relatively rare. They formed in the very earliest moments of the solar system 4.5 billion years ago.

So suddenly, instead of considering quasicrystals formed in some laboratory, we're talking about quasicrystals that formed in nature even before there **was** an earth, even before most of the minerals we know of were first formed.

Moreover, one of the world's experts on this famous class of meteorites was the guy from the Smithsonian we had been talking to all along, Glenn MacPherson. I won't go through all the stories of the various questionable emails along the way about why we were even looking at this, but suddenly things had shifted. *This* was the definitive test, and there was no question that it was correct. Suddenly this was really interesting to him, and we had to do a lot more tests. The only problem was there was no more material to test—we were out of test material.

And since he was telling us that there was nothing else like it, the only hope was to go back and look for more. Most people—99%, maybe even 100% of geologists—would have said, "*Just because you found a grain here in 1979, doesn't mean you're going to come back over 30 years later and find something again. The chances are it could have been an air burst or something like that. You'll never find another grain of this stuff.*"

It seemed like a crazy mission. Furthermore, it's a restricted region of Russia. It's not like you can just buy a ticket and go there and look. You need permissions from the Russian government, the FSB —the modern KGB—the military, the local government of Chukotka, you need transportation to go out there... Just think about the pile of things you need to do to go out there.

HB: But you're not going to stop now, surely.

PS: You also need money to make this work. But I felt that if we didn't go then, we'd never be able to go, because Valery was willing to go to help us, and he's now in his 60s.

Within eight months of deciding to go, we were on the ground there. We managed to get enough money together to support the trip. Vadim Distler and his colleague Marina Yudovskaya formed a team which helped us make the local and governmental arrangements, and all those things we had to do. We brought them to Princeton at one point because it wasn't clear how we were actually going to run this mission.

HB: OK. But let's skip all that stuff. I want to know what happened.

Questions for Discussion:

1. Do you think that Paul would have been able to make any real progress on his khatyrkite quest in an age before the invention of the Internet?

2. Why do you think that Razin mentioned Valery Kryachko by name on the holotype paper if he wasn't prepared to let him share authorship?

3. Why do you think so many people Paul encountered were so willing to help him in his quest? What does this say about the universal appeal of the scientific enterprise?

IX. Kamchatka

Closure, and perhaps another beginning

HB: So what happens? Did you find the stream? Tell me.

PS: Well, I should say one thing first: when we were originally going to go, my notion was that I was not going to go, because I'm not a field geologist. Actually, I've never even been camping before, so this hardly seemed the place to have my first experience of that kind. And I'd probably do something stupid like break a leg or something and cause them to end the mission, so I had a lot of concerns about that.

I had this great plan where we'd helicopter people in to the site and I'd stay back in the town of Anadyr, while a few people would go and do the actual work. But it turned out that by the time Vadim and Marina came to visit Princeton the helicopter idea wasn't going to work. We just couldn't get helicopters or proper insurance.

So we decided that they would go in trucks, and they said, "*You must go, Paul, because there's now enough room to go, and you really ought to.*" So I was sort of dragooned into going. I remember telling them, "*I've been looking at Google Earth and there are no roads.*" But they told me, "*No, no. Don't worry. There are roads.*"

So we get off the plane in Anadyr after a long trip with our whole team, a combination of Americans and Russians. There were also some graduate students who would apply some much-needed muscle power, among them my son, Will, who was a geoscientist at Caltech, and is now a geoscientist at Harvard. That was a great aspect of it for me personally.

So we get to the place where the trucks are, and these are not conventional trucks. These are treaded, tank-like vehicles: the base was like a tank and the top was like a beat-up van. So those were our vehicles. The next day we're out at the edge of Anadyr ready to set off in the trucks on this permafrost tundra, which you can't really walk on very smoothly, and which these trucks can barely manage to make their way across.

We were moving across at a snail's pace for four days. Every day we were going up, down and through streams. We had some near-fires and broken axles and a bunch of adventures along the way. We encountered bears and various other challenges. But we got there.

We spent the first afternoon looking for the spot where Valery first found the sample more than thirty years ago. Of course, things have grown since then, but after a few hours, we actually found the spot. Then began 10 days of hard work, digging material out—not just there, but up and down the stream. You dig it out and then you have to pan it, much like you would pan for gold, except Russian-style. Russian-style panning is not like the Western style of panning. It has its own protocol. Valery was a world-expert panner. He loved to pan. Will loved to dig, so he was the main digger.

We dug about one-and-a-half tons of material out of there, panned it, and carefully labeled it so we knew where everything came from. We had one team that was structurally mapping the geology of the area in case it wasn't a meteorite; and if it was a meteorite, to try to understand better how it had fallen.

We also had a little laboratory in which we did some analysis. But you can't tell from anything you see in the field whether you've got a quasicrystal or not, or even if you have the right chemistry —for that, you need expensive instruments. But you *can* do some very rough, visual things. So we did the best we could to help guide matters with what we had.

Ten days later, we packed up and left with lots of samples, not knowing what we had. By the time we left the mountains in early August, winter had come to Kamchatka—snow was falling—so we barely got in with enough time for things to have melted and barely got out before winter came: it was a perfectly-timed mission.

There were also some issues with how to get the material out of Russia, but we eventually managed to do that as well.

Then came the tedious task of looking at the material grain by grain. It's like looking at sand grains one by one—and you have many grains of sand—which, by the way, are still sitting in boxes in this room, because we're still looking through them. Luca was mainly the one in charge of all that because he knew what to look for.

And six weeks later, he found a sample that looked like meteoritic material, took its X-ray diffraction pattern, and it turned out to be icosahedral quasicrystal.

After all this time, after this crazy story which I told you that you could have—and should have—been sceptical about—and even *we* were sceptical about—suddenly, we had actually managed to go there ourselves, pick something out, bring it back, and find quasicrystal in it.

That was a key moment.

Nobody expected to find anything, but the fact that we did manage to find a natural quasicrystal, just verified everything in the story that we had been putting together up to that point.

Since that time we've collected something like nine grains of material of the same meteorite. The meteorite was very heterogeneous—different things happened to different parts of it—but there are enough commonalities that you can tell it's part of the same thing.

They all have either khatyrkite, or cupalite, or what's now called icosahedrite—the first natural quasicrystal—in it. Then comes the challenge of analyzing what other kinds of novel things are in the grains and how the thing formed. That's what we're in the midst of doing now.

In the process of doing that, we've discovered a lot of interesting things that are unique to this meteorite. We discovered that these high-pressure effects we saw actually repeated again in various parts of the sample, which means it underwent a very high-velocity impact—much higher than has been observed in any CV3 carbonaceous chondrite before, which might have something to do with the story of how it formed.

We've also found other new mineral phases, most of them crystalline—and recently we officially announced the discovery of the second quasicrystal we've discovered in nature, from the same meteorite.

It's a different material—aluminum, iron, and nickel, instead of aluminum, iron, and copper—and a different kind of quasicrystal. It's not icosahedral, like the sample we were discussing. It's what we call decagonal—it's ten-fold symmetric in a plane: stacks of ten-fold symmetry, like Penrose tilings stacked one on top of the other. It's a completely different symmetry and structure.

So we've found two different quasicrystals in the same meteorite, as well as other interesting things. There are a lot of different tests that are happening right now to explore other issues.

Just to name some of them: with a group in Zurich, we're trying to extract isotopes of helium and neon from some of our samples, which preserve information about the size and timing of objects undergoing a major impact.

In another experiment, at Argonne Labs, we're actually reproducing the high-pressure conditions for synthetic materials of the same nature, and we're asking if the quasicrystal is stable when you put it under ultra-high pressures. We put it under ultra-high pressure and study how it transforms, or doesn't transform, as a function of temperature and pressure. That way, we can see if pressure helped or hindered the formation of quasicrystals. It looks like it helps it.

And with a group at Caltech we want to reproduce the collisions at these high velocities. We think we have some ideas as to what the sequence of events was, but we're trying to actually reproduce it in the laboratory, to some degree, to see if we can figure out how you get metallic aluminum, together with copper, all in one material. That's a big puzzle.

If you were to ask most petrologists and geophysicists, they'd tell you, "*That's the part we **really** don't understand*". We have some clues as to what some of the aspects were, but unless we can actually reproduce it in the laboratory, we can't be sure that we've got it right. We're trying to do that now. All of these efforts and more are continuing as we speak.

HB: As I mentioned to you earlier, Freeman Dyson urged me to talk to you saying that you had a really amazing adventure story, but I must admit that I thought he was exaggerating. But this is really an Indiana Jones type of story.

PS: It turned out to be. You feel really lucky to be in such a story, I have to say. At some points, though, you don't feel so lucky, like when you go a certain distance and then get stopped. But the fact that we managed to get ourselves through all those points...

HB: ...and then winding up with this meteorite. I wanted to ask you about this particular type of meteorite you are dealing with. I guess one obvious question involves comparing it to other ones that are out there. Presumably you've looked at comparing samples from others.

PS: We haven't really done that yet because our focus has really been understanding this one sample, and you can see how much work that takes.

But I've been discussing with various people how you might survey lots of such materials to look for samples. The challenge is that there are some processes that you might be able to use, but they'd be destructive; so you might discover that you had found something but then you wouldn't have anything left. So I'm exploring strategies for looking at large bulks of material, or I'm hoping that people just looking at their own favourite samples might now join in the search and find more of these materials.

I'm also interested in continuing the search for quasicrystals generally. We've now found the first and the second natural quasi-crystal. And if you have one, you'd think you'd have two, three, four or more, so I'm trying to find some other ones.

One of the reasons why I was originally interested in the search was not just because I wanted to learn something about the stability of quasicrystals. I was hoping that, rather than using serendipity in the lab to find quasicrystals, there might be serendipity in nature to find new quasicrystals. So far the two examples we've found are not serendipitous in the sense that they are actually ones that we already knew existed in the laboratory.

That's not a complete coincidence: we were using what we knew as information for what things to look for. I want to think about how to continue the search in a way that we might find new materials that we could then bring back to the laboratory to try to synthesize rather than the other way around.

That wouldn't just be a useful way to expand the catalogue of quasicrystals, you might also find examples with other physical properties which, combined with the quasicrystallinity, would make them very interesting materials.

Questions for Discussion:

1. Are you surprised that this story is not better known? (For many more fascinating details of this remarkable story, readers are strongly recommended to read Paul's 2019 book, **The Second Kind of Impossible: The Extraordinary Quest for a New Form of Matter** published by Simon & Schuster.)

2. Would you have gone along on the expedition to Kamchatka if you had been able to do so?

X. Passing It On

How to keep the flame of science burning brightly

HB: Wow. It seems to me that every single person who thinks that "science is boring" should experience this story. Which brings me directly to the following question: what advice would you have for young students when it comes to pursuing a scientific career?

PS: I think the most important thing is to get some experience with research. Classroom teaching is very different than the real world of scientific research. In classroom teaching you're taught things that we understand well enough, things that can be brought to a level that many people can understand. But when you're in research, you're really on the frontier. You don't know which way to turn or which way to go. You have to decide that for yourself.

That's extremely exciting, I think, for those who are scientifically-inclined. You can even do research at home. When I was a kid, I had certain projects that I could work on at home. They may not have been very important research projects, but that wasn't the point: I was enjoying the sense of discovery.

Later on, as soon as I had the opportunity, I had the chance to work in laboratories in my community. If you have access to that, I think you should try to get to that at as early an age as possible. Unfortunately it's harder for kids to do that today—at least in the US—than when I was kid. I was able to get into a laboratory, I think, at age 12. I had to knock on many doors, but I was finally able to get in the laboratory.

But nowadays, on the Princeton campus, say, there are actually work laws in the books that prevent young people from going into a laboratory until they're 16, which is pretty late, I think, for getting started. The issue has to do with safety, but we let 12-year-old kids play football, while we don't let them into a scientific laboratory to work, even under relatively safe conditions. We keep them from that kind of activity, and I think that's really bad.

I've been thinking about how one can change that regulation, because I think it's bad for the future of science. We lose people that way, because they're essentially being told, *"This is bad, go do something else."*

HB: Like football.

PS: That's right. I have nothing against football, but I think science is also an exciting activity—and getting that opportunity, for me, was really important.

I think that this is especially true when you're talking about bright kids, many of whom are bored in school and could really benefit from this kind of activity, which makes subjects much more relevant to them. I think that improving the interface between schools and research labs, finding a way to change these regulations, is very important for the future of young scientists.

HB: Picking up on your comment about some of these students being bored in school, what advice might you give to a teacher in terms of how to keep kids a little more stimulated when they are at school?

PS: I think it's a challenge for teachers. Some teachers feel that they're supposed to be the experts and they're supposed to know everything, so when they get these very bright kids coming at them questioning things, they may react badly because they feel like their authority is being challenged.

But I think a healthier attitude is to recognize that it's perfectly okay to say, *"I don't know."* I do that all the time. A much better response is, *"That's a good question. We should pursue it,"* encouraging students to ask difficult questions, reinforcing their curiosity.

It's also important to follow through. It's not enough just to say, *"Oh, that's an interesting question!"* It should be instead, *"Let's pursue this. Let's figure out how to pursue it. Let's figure out how we might answer that question."*

We're not living on an island, especially with the Internet. If you have a question about something, go find the expert in it.

When I was a student, the first project I worked on was a math project. I didn't know at the time, but it turned out that it was written by a famous mathematician. I remember writing to him with my little calculations that I was doing as an elementary school student and I got a wonderful letter back from him that was not only very supportive but also helped me work out something I was doing.

You have to appreciate that many people are willing to help and support in this way, so take advantage of that fact to help those students see the bigger world and interact with it.

You never know where it's going to lead. There are also all kinds of opportunities for bright young kids to spend summers in various kinds of camps and engage in research. Some people know about them, but many people don't. It's important for teachers to investigate what opportunities exist for those bright young students.

The nice thing about those activities is that not only do they give research or advanced-learning opportunities, they also provide occasions to meet other similar types of kids. Very often, especially in smaller schools, these very bright kids feel uncomfortable: socially different and isolated. And by participating in these programs, they discover that they're not alone, that there are other kids like them out there.

My kids were involved in activities like that and they've made life-long friends that they continue to interact with from those activities.

In short I think there's lots that can be done by capitalizing on what's available and accessible. When I was a kid, it was very important to live exactly in the community where the lab was, but nowadays it's different: you can work with satellite data that everyone has access to and interact with a scientist who's on the other side of the world. There are many things you can do. I think that it's important to be creative about how to use those resources.

HB: Very wise words. Thanks a lot Paul. This was wonderful.

PS: Great—you're welcome.

Questions for Discussion:

1. Do you agree with Paul that students should be encouraged to participate in research activities at an earlier age? Is there a way to do so while ensuring appropriate safety measures?

2. Why do you think teachers often feel compelled to refrain from admitting their ignorance about a specific topic? Is there any way to change such attitudes in a scalable way?

3. Has this conversation made you appreciate that some areas of science that you previously thought were uninteresting might be worth learning about more?

Continuing the Conversation

Readers who enjoyed this conversation are strongly encouraged to read Paul's memoir, *The Second Kind of Impossible: The Extraordinary Quest for a New Form of Matter*, published after this conversation.

Cryptoreality

A conversation with Artur Ekert

Introduction

Putting the Pieces Together

For centuries, what we now know as "physics" was universally regarded as "natural philosophy". For those who might be inclined to conclude that this is simply a case of nomenclature, it is useful to look even further back and appreciate that some of the oldest philosophers of the Western philosophical tradition—people like Thales, Anixamander, Anaxagoras, Leucippus, Democritus and many more—were nothing less than the physicists of their day, devoting the lion's share of their attention to developing theories to account for what, exactly, the world was made of.

Aristotle famously referred to such thinkers as *physiologoi* or *physikoi*—those concerned with developing models of nature based on purely physical principles, as opposed to the *theologoi* and *mythologoi* whose explanations necessarily required the involvement of and intervention of the gods.

In many ways, Isaac Newton represented the culmination of this tradition. His *Philosophiae Naturalis Principia Mathematica,* published in 1687, was so widely recognized as the optimal marriage of mathematical rigour and critical philosophical inquiry that it convinced many people that this was the *only* bona fide intellectual approach to increasing our understanding of *anything*.

Well, that's a very long story that we're still very much in the middle of, but one of the clearly identifiable and somewhat ironic effects of Newton's monumental accomplishments is that his efforts were *so* successful and *so* dominant that they went a considerable distance towards separating philosophy from the natural sciences.

Since there no longer seemed much point in debating between different approaches to understand how the tides worked or the planets moved, those of a more philosophical persuasion tended to increasingly refocus their attention on more human-centred issues like aesthetics, ethics and political theory, while those interested in working out the finer details of ballistics or optics found themselves firmly ensconced in the emerging technical discipline of "physiks".

And as the centuries ticked by, there was ample evidence for physicists to convince themselves that the steady paring away of their philosophical origins was all to the good. From the Industrial Revolution spawned directly by Newton's powerful insights to Maxwell's stunning unification of electricity and magnetism that gave birth to the modern communications age, the physical sciences moved steadily, rigorously onward, penetrating ever deeper into nature's secrets and rapidly transforming societies in its wake.

To a physicist, philosophy was viewed as no more directly relevant than linguistics or art history, with any special bond between the two of mere historical significance springing from a time when our inquiries of the physical world was in its childhood, a condescending testament to how much progress physicists had made over the centuries—and how little their philosophical colleagues had.

This sense of smug complacency was tested in the beginning of the 20th century when quantum mechanics forced scientists to confront the previously unimaginable prospect of a rigorous theoretical framework of nature that seemed impossible to actually comprehend, but their confidence eventually returned a few decades later when it became apparent that, for all its befuddling incomprehensibility, the theory was nothing less than wildly successful in predicting how nature would behave.

Indeed, in time, the triumph of quantum theory only convinced physicists that paying attention to philosophical issues was to be avoided at all costs.

"Do not keep saying to yourself, if you can possibly avoid it, 'But how can it be like that?'" cautioned the celebrated American physicist Richard Feynman, *"because you will get 'down the drain' into a blind alley from which nobody has yet escaped. Nobody knows how it can be like that."*

And so, for generations of physicists, philosophy was no longer seen as something quaint and irrelevant, but nothing less than the *enemy*: a dangerous temptress that, if unchecked, would impede our ability to make future progress. Thinking deeply was henceforth to be avoided. *"Shut up and calculate!"* would become the new motto.

There were some, of course, who never submitted to this new dogma, most famously Einstein and Schrödinger, two particularly cultivated, philosophically astute giants of 20th-century physics who had done so much to establish quantum theory to begin with. Both had worked diligently and tirelessly to develop penetrating thought experiments to expose what they saw as inherent philosophical contradictions associated with the theory. Schrödinger, among other things, developed his famous "Schrödinger's Cat" scenario, while Einstein's objects to the philosophical implications and perceived contradictions of quantum mechanics were perhaps most pointedly attacked in a 1935 paper he wrote with Boris Podolsky and Nathan Rosen generally referred to as "Einstein-Podolsky-Rosen", or just "EPR" for short.

EPR demonstrated, in characteristically penetrating and lucid prose, the profound incompatibility of the theoretical framework of quantum mechanics with our commonly recognized assumptions about reality, explicitly highlighting where this incompatibility lay in an effort to demonstrate the inherent incompleteness of quantum theory. Formally, the argument went, *Either we accept that reality is this way or quantum theory must not be the whole answer.* Less formally the implication was clearly, *Everyone knows that reality must be this way, so there must be something missing with quantum theory.*

It was a brilliant, customarily Einsteinian work of "theoretical philosophy"—and it was, given the stridently anti-philosophy zeitgeist described above, almost universally ignored. In fact, despite Einstein's reputation as the greatest theoretical physicist of the age—if not, indeed, in history—it took almost three decades before any significant work was done on EPR. And when it did finally happen, it came from a particularly curious quarter indeed—John Bell, a relatively unknown accelerator physicist who was simply trying to make some sense of what it all meant in his spare time. Remarkably, Bell was able to push the conceptual ball forward considerably, rephrasing the original EPR argument in a way that it could be clearly subjected to a demonstrable experimental test in a seminal paper of 1964.

And then, yet again, nothing.

It took another decade or so before a few people tried to do the experiment, with contradictory results. Finally, in the early 1980s, Alain Aspect conducted a comprehensive series of experiments that settled the score once and for all in a way that would clearly have shocked Einstein, Podolsky and Rosen: quantum mechanics, it seemed, was fine. It was our basic assumptions about reality that needed an overhaul.

At this point, you must be saying to yourself, **something** must have happened. Surely such a result would have done nothing less than send shock waves rippling throughout both the entire physics and philosophy communities. But once again—inexplicably—no. Not really. The physicists had shut up and were all quietly calculating things. And the philosophers were doing—well, whatever it is that philosophers do.

It was left to a young, Polish graduate student at Oxford ten years after that, in the early 1990s, to demonstrate how these remarkable ideas could be further harnessed in an intriguingly tangible way—developing a rigorous cryptographic scheme for protecting information flow based upon a detailed appreciation of EPR's "theoretical philosophy".

"I was simply driven by philosophical curiosity, saying to myself, 'Well, if something doesn't exist, there's no element of reality; and if information is physical and thus encoded in something, and you've

encoded it in something that is, somehow, losing its element of reality, then there's nothing to eavesdrop, because it doesn't exist.'"

Were you to do a poll, within universities or the public at large, you would be hard pressed to come up with two disciplines that seem further apart than metaphysics and cutting-edge surveillance systems, even to the most fanatical advocate of "interdisciplinarity".

But they're not.

I wonder how long it's going to take to appreciate *that*.

The Conversation

I. Beginnings

Mathematics, physics and intuition

HB: I'd like to start with your scientific beginnings. Did you think you were going to be a professional scientist when you were a little boy?

AE: Actually, I'm not sure. It's certainly not the case that I was always exclusively focused on science. You know, you go through these different stages—you want to be a policeman, you want to work for the fire brigade, then you want to be a film star. So, I think I went through all of this and then at some point two or three books made an impact on me. One of them certainly was *The Feynman Lectures on Physics*. That's probably a cliché, I realize—probably many guys in the field tell the same story.

HB: Well, if it's true, it's true.

AE: You know, like him or not, Feynman certainly had a big impact on me, a certain degree of clarity while showing that you can have both intuition and mathematics coming together nicely in order to understand something about the world out there. It was very good, it was quite impressive.

Another book, which I liked a lot was *The First Three Minutes* by Steven Weinberg. What is particularly amazing about that is that at some point you appreciate that you can learn about something that is not given to your senses: we were not there when the big bang occurred, but somehow, in your mind, you can reconstruct with a certain degree of certainty what happened.

So it was really a sense of the power of science, the power of scientific methodology: we can develop these bold conjectures and then sometimes refute them and come up with other conjectures, somehow all the while giving us a much broader perspective on the world. Isn't that amazing?

HB: It's remarkable. How old were you when you were first exposed to these ideas?

AE: I think I was about 16 or so.

HB: Was there any particular orientation towards science in your upbringing and family environment, or did you just stumble upon it by accident to a certain extent?

AE: I think I was certainly encouraged to do science, but there was no pushing in any direction. In fact for a long period of time, I was much more interested in all kinds of sporting activities.

HB: Which sports?

AE: Well, just about everything, I did everything. I'm one of those characters who played football, hockey, swimming, I did judo for a long period of time. I'm a very outdoorsy kind of person. I'm not so much an urban person. I like to pop into town every now and then and benefit from cultural life and interesting cafés and so on, but generally speaking I feel much better outdoors. I'm really a country boy deep down.

HB: And there's also the scuba diving, which I know is a big deal for you.

AE: Yes, I do scuba diving, I fly light airplanes, I do all these kinds of crazy things but, you know, some people say I do it because I like risky things, but it's not so much about taking a risk; it's more about trying to concentrate on something, and this is the best way to forget about some other things.

If you do things that are a little bit on the edge, an activity that requires a certain degree of attention—be it scuba diving or flying airplanes or wakeboarding or whatever—then it's very relaxing because you focus on something, and then you don't have to think about other things, like some mathematical problems you're working on.

Every now and then you need to reformat your brain; and for me the best way to do it is to do something that I have to devote all my attention to. This usually can only happen when you do something that is critical: that if you don't do it properly, then you die. So, there's no special predilection for risky things. I may be adventurous, but I'm certainly not trying to take unnecessary risks.

HB: Do you find that when you're focusing on something like this— when you're flying a plane or scuba diving or what have you—by focusing your mind in a different direction, does it sometimes lead to breakthroughs in understanding, in the same way that sometimes people find the solution to a problem in their dreams?

Do you ever come back after scuba diving or flying a plane and find that you have a slightly different, more productive, perspective to a physics problem?

AE: Sometimes after, but not during. There's this moment of concentration and then you relax.

Many people have these stories about good ideas that they've had when they were in the shower or something. That's never really happened to me, but what is certainly true is that this period of time *after* you've been concentrating on necessary procedural things associated with diving or scuba diving, when you can relax and engage in lateral thinking and let your brain make all these kinds of random connections, *that* can be very productive.

Sometimes you don't even realize that it's a productive period of time, but it is: your brain goes into some kind of a different mode and many good ideas that I have happen when, after a period of concentration, I was just relaxed and tried to do nothing.

Maybe that's a good thing: trying to do nothing rather than something. Sometimes doing nothing can be really creative, right?

HB: I think so. Bertrand Russell wrote a typical thoughtful and insightful essay called *In Praise of Idleness*, and quite a few others have held forth on a similar theme.

AE: Yes. And in Japanese, there's a classic book called *Essays in Idleness: The Tsurezuregusa of Kenkō*. I also think that there are some aspects of idleness or even laziness, which are definitely conducive to creativity, like looking for shortcuts.

HB: Right. It's also worth mentioning that it's not so easy these days, being idle. It's often difficult to get people to sit still for a sufficiently long period of time that would constitute idleness—they're constantly looking at this or that.

AE: Yes, you're right. I don't know what it is. Maybe because our attention span is not what it used to be with the internet and everything—you try to dose knowledge in some kind of digestible quanta to keep your attention so that it lasts no longer than a few minutes—perhaps even less than that, actually—since everyone is so rapidly jumping or clicking from one thing to another.

HB: Right. But—he says, rapidly changing the subject—let's get back to your story. You've read a couple of interesting and motivating books such as *The Feynman Lectures* and Weinberg's *The First Three Minutes*.

Now you're in high school. What happens then? Do you start thinking more about science as a career at some point? First of all, are you primarily focused on physics at this point? Presumably that's the case, given the two books you mentioned, but perhaps not.

AE: I would say more mathematics than physics; I was (and am still) much more attracted to mathematics, but it was much more recreational mathematics. I really like those mathematical puzzles where you have to find this nice solution to a problem.

That is certainly something I still do, especially on long flights when it's often not so convenient to read another book. The great thing is to pick up a problem—a simple problem, a mathematical problem, one of those nice, recreational problems—and you try to solve it. There are two options: either you solve it and then find another one, or you fall asleep, which is equally good on a long flight.

I love those mathematical problems. So mathematics was certainly the first thing, however, physics seemed to be much more mysterious, because in mathematics you think, *We set up the rules of the game. We can design a formal system and try to prove things within this formal system.*

It's very interesting that this formal system is usually, somehow, related to what is going on in the real world—but that's another story, maybe we'll get to that later—

HB: I expect so.

AE: But, at least at some superficial level, you reach the following conclusion: *In mathematics, we set up the rules of the game, but in physics, they are set up for us and it is our job to discover them.* And at some point I remember thinking to myself, *It might be a bit boring to play by your own rules all the time, why don't you try to play by someone else's rules and, at least, try to find out what they are in the first place?* So that brought me to physics. But I was always oscillating between physics and mathematics.

Somehow, I've been lucky in my career to be able to do things that brought together computer science and mathematics, through topics like cryptography or secure communication, while also bringing in many beautiful ideas from quantum physics. So that was luck, I think.

HB: You did your undergraduate degree in Poland, right?

AE: Yes.

HB: And then, you decided to go to Oxford to do your PhD. I have some sense of what came out of your PhD, but I don't know how you felt going into your PhD or what you thought you were going to do. What originally attracted you to Oxford and to that whole course of study?

AE: You know, at some points in life, you don't know what you really want to do. For me, doing a PhD was essentially a way to prolong this period of time, thinking to myself, *Okay, I'll have a few more years to decide what I really want to do.*

I didn't necessarily anticipate at the time that I would end up doing physics as a career. I still thought that it was a bit too boring, because I felt that doing that would confine me to offices and lecture halls for the rest of my life, and that was not very appealing to me.

To be honest, somehow, until Oxford, I didn't find that many role models, people that I was really taken by—there was no direct evidence in my life that there were "non-boring" physicists, for example, fascinating personalities that I would like to talk to and discuss things with.

That changed in Oxford, somehow. I met David Deutsch, who made quite an impact on me, he was my co-supervisor. And Roger Penrose also certainly made an impact on me. And I had the chance to learn quantum optics from another person, Peter Knight, at Imperial College. I visited him a lot and that was a very good experience.

So all of a sudden, I ended up in a place, which was completely dysfunctional otherwise—it is, yes, feel free to quote me on this—

HB: I don't have to quote you: I can just play the video. You just said it on camera.

AE: Right. So Oxford *is* dysfunctional, but this also good in many ways because it gives you the freedom to do what you want. What I like about Oxford is that there is no centre of power: the decisions are completely diluted and the system is not very coherent. It takes a while to make any decision going in any direction, which is both good and bad.

When decisions have to be made fast and you have to be decisive—be it some funding opportunity or something—then Oxford is hopeless, I think.

But when it comes to arriving at some sort of long-term vision for something, then it usually works. And the concentration of very talented people that Oxford attracts—often real characters, eccentrics—is wonderful, because the more you have, the greater the chances that more will come and join the club.

So, I suppose it's a cliché to say this because it's obvious, but the quality of Oxford lies in the quality of people who are there.

HB: And while there's certainly something to be said—as you just did (and we have it on camera)—about being able to act quickly and coherently to capitalize on funding opportunities or what have you, there's also something to be said—getting back to your earlier comment about idleness—for *not* acting, leaving people alone to get on with doing what they want to do: thinking, creating and so forth.

AE: Yes, exactly. That's what happened in my case; I never felt that Oxford pushed me in any direction, but it offered so many opportunities to talk to interesting people. It was, at the end of the day, my decision what I wanted to do.

Of course there were people who were supposed to talk to me every now and then, but in the end the message was clearly, *You are in charge of your own career: if you want to do something, do it; if you don't want to do it, don't do it.*

But I really enjoyed working with David Deutsch. He had many great insights, and he changed my way of thinking about physics.

HB: How so?

AE: Well, actually, when we met for the first time, we didn't talk physics at all. I don't even know exactly what the subject was but we ended up talking about Karl Popper because we realized that we were both Popperians. David was very much into Karl Popper, not only his scientific methodology and how science works, but also his broader societal views as described in *The Open Society and Its Enemies*.

For me, living in Poland for such a long period of time, I thought that this book was fantastic; it really showed the dangers of any extreme version of politics, be it communism or fascism, as well as Popper's view on democracy and elections—views that are still, I think, very important today.

So David and I talked a lot about this, and somehow, we clicked. When it came to physics, we talked mostly about quantum mechanics. I hadn't thought that much about interpretations of quantum theory before meeting David. I knew that there was a bit of an issue there, but I didn't realize how important it was.

As you probably know, David is very much into the Everett interpretation, which seems pretty bizarre to most people the first time they hear about it.

HB: Sometimes the second time too.

AE: That's right—or the third time, for that matter. My gradual conversion to the Everett interpretation was a fascinating process. I felt like a religious convert: you gradually start adopting this worldview that comes together as a package. David never persuaded me or wanted to push me in that direction—*This is the dogma; this is what you must believe.*

It just came as a package. We talked about quantum computing and some other ideas—and then, somehow, it was always easier or better to talk about it when you looked at it from the perspective of the Everett interpretation.

So that was one thing that happened. Philosophy was always of some interest to me, but somehow I escaped thinking too much about the interpretations of quantum mechanics in the early stages of my career. As an undergraduate, I didn't pay too much attention.

HB: It's probably healthier that way.

AE: Well, maybe. I think that what often happens is that at an early stage of your career you judge the difficulty of the subject and the mathematical tools without fully appreciating the conceptual part.

Of course, certain mathematical tools are much more sophisticated than others, but at the end of the day, anyone can pick up mathematical tools. But somehow it's not enough just to learn mathematics to do good physics. At some point you also have to use a bit of intuition. Where does it come from? I honestly have no idea honestly, don't ask me. Even after a few bottles of wine, I'm still not sure where it comes from.

HB: Well, there's also the question, I suppose, of what we're talking about. What do we even mean by mathematical or physical intuition? Perhaps asking, *Where does it come from?* is just another way of asking, *What is it?* I'm not sure.

AE: Yes, there are many questions there, of course. There's this essential question, *This mathematical world, is that something you discover or you invent? Does it have some kind of existence on its own?*

Personally, I would assume that it does because those mathematical concepts reflect such a shared experience, and we usually believe that a shared experience comes from something that has its own independent existence.

Regardless, if you take mathematics at instrumental levels—as I was sayings, of tools—and you want to use them in an efficient way for doing physics, it's not enough to be fluent with using those mathematical tools; there's something else, another ingredient.

What it is exactly, I don't know: it's a bit of intuition; and sometimes I even think that being too skillful with those mathematical tools is not necessarily of great help—maybe it's better to have this intuition to help build your models and deepen your understanding.

Anyway, I tend to think in very geometric ways—my mathematics is basically geometry—but what you were asking is, "*Where are those tools coming from in the first place?*"

Which prompts the question, *Can we think about something that we do not, somehow, abstract from our perception of reality?* Maybe those mathematical tools are just a very high form of abstraction of what we already see and that's why they are so useful in describing reality. That might be the case.

HB: I would very much like to get back to some of this later on, but first I'd like to finish off this section about your personal background. Another point worth highlighting about you is not only your deep knowledge of physics and mathematics informed by various philosophical tendencies and orientations, but also your considerable experiences on the "human side", as it were, having built and managed numerous groups of scientists, dealing with all sorts of different personalities and human dynamics.

In my experience, there tends to be a sort of dividing line between types of researchers, at least in theoretical physics. There are those who are very formal and very mathematically-oriented, and then there are those who have this sort of intuition, as you briefly alluded to. We were talking before about mathematical intuition, but of course intuition comes in various different flavours—and here I'm referring to a more physical-oriented type of intuition, a sort of Feynman-like sense of the world.

Is that your experience as well, that you can roughly divide people into these two categories? Or is it more complicated than that?

AE: Yes. They are not disjoint sets, I would say, but overall yes, I think you can. There are those who think in very abstract terms—and they can be very successful, of course—and those who are driven by intuition.

Some people can see the solution before they do all the mathematics, but intuition can be misleading too, of course: in many aspects of mathematics, you might just end up in a cul-de-sac because you used the wrong kind of intuition. But I think that, in most cases, it works.

However, I do agree that there are those who think like physicists and then there are those who think a little bit more like a mathematician. There's a very good puzzle that you can give to a student that highlights the difference between the two—

HB: Hang on: I'm going to cut you off right now. This is the one with the light bulbs, right?

AE: Yes.

HB: I know this is going to sound extremely regimented, but I'd like to fit this in later, if you don't mind. But I promise you, we'll get back to it.

AE: OK.

Questions for Discussion:

1. To what extent do you think that mathematical or physical intuition can rigorously be defined?

2. Why do you think Artur begins to talk about mathematical Platonism shortly after he mentions the concept of mathematical intuition? What is the link there? Readers particularly interested in this topic are referred to the Ideas Roadshow conversation **Plato's Heaven: A User's Guide** *with philosopher of science James Robert Brown.*

3. Are you surprised at the idea that some physicists would also feel a resonance with Karl Popper's non-scientific views in addition to his opinions on the methodology of science? Should people like Karl Popper be presented as systematic thinkers rather than in terms of specific, compartmentalized views?

II. Cryptographic Essentials

From the ancient Greeks to the Cold War

HB: Let's talk about cryptography now. Maybe I can start by asking you to give a very brief rundown, as you've done many times before, on some of the basic issues in cryptography from ancient history up to the present day, in terms of the concepts of what people were trying to do leading up to the notion of a one-time pad.

AE: Sure. First of all, we talk about this umbrella of cryptography, there are many different tasks, but let's talk about the simplest ones, which means two individuals, like you and me, where the task is to communicate in such a way that no third party can eavesdrop on this communication. This is the most common, classical scenario in cryptography—there are others, of course like password protection, authentication and so on and so forth, which are topics on their own.

Within this simple scenario what people tried to achieve was to make sure that the sender could send a message to the receiver in such a way that, even if it is physically intercepted, it looks like gibberish, it doesn't look like something that you can decipher.

We know that the origins of secret written communication basically goes back to the origin of the alphabet, which is quite interesting, because—while I don't know what historians really think about this—it seems to me that it's a reasonable conjecture to assume that the alphabet itself was very important.

In China, for example, there's no record of sophisticated cryptography, playing with so many characters was probably too difficult. But if you have an alphabet that has 26 or 30 characters to play with, then you can do it: you can permute them, you can replace one character with another and so forth. There are many relatively straightforward ways that you can scramble a message that is made out of 26 or 30 characters.

HB: That's fascinating in its own right, because it seems to be another, rather distinct, way of how language can influence thought. If it's the case that having a language of 26 characters gives one a natural predilection—or at least statistically encourages people to start thinking in a particularly cryptographic way—that's suggestive, perhaps, of something. I don't know, I'm just speculating.

AE: Well, to the best of my knowledge, it's not that secure communication was unknown in ancient China, it was just simply done differently: people would rather hide messages in mechanical devices.

HB: Sure, I'm not saying that all Chinese people are any more or less trustworthy than anyone else, I'm just speculating that this might be another revealing example of how language can influence thought in some particular way.

AE: Yes, I think it is. The way you encode ideas in terms of graphic characters is important, because then what you can do with those characters then shapes the way you think about things. That seems to be true.

Certainly what is true is that the origins of cryptography, as we know it today, go back to ancient Greeks and ancient Romans; and there's not so much we can learn about it from ancient China, for a number of reasons—partly because there are not so many documents that have survived. We know about some ancient Chinese poetry where some messages were hidden in a certain way, but nothing at the level of sophistication that was developed in Europe.

The way this was typically done in Europe involved what we call the *permutation* of characters of the alphabet: you'd scramble a message by writing it out and then moving the characters in the message so "A" goes to "D", say, "B" goes to "W", and so on.

But, of course, if I do it, then you also have to know how I did it in order to unscramble it. And one of the first ways to do this, was done in ancient Greece by using a device called a scytale, a wooden baton.

The idea was that I would take a strip of parchment and wrap it around this baton and then write a message on it lengthwise. Then I would unwrap it; and the result would be a strip of parchment with characters all over the place, making no sense. Then, I would give it to a courier to take the parchment from me to you.

Now, suppose it was intercepted—it would be very difficult to figure out what the message was. But meanwhile, you would have a wooden baton of the same diameter—in other words, there is a secret we would share: you would have a wooden baton of exactly the same shape, so that once you received the parchment you could simply rewrap it around your baton and the message would reappear.

A scytale encoding "Mary had a little lamb..."

That was the first mechanical device that we know about that implemented the permutation of characters for cryptographic purposes. Then, somewhat later, ancient Romans did similar things—Julius Caesar is alleged to have come up with a cipher where he would consider substituting one letter with another letter—but again, if we want to use this method, we have to agree on how we are going to do it.

So, one way to do it is to say, "*Let's just shift the character.*" In other words, we define a certain rule of substitution by shifting the alphabet by a certain number of characters. So, for example, I take "A, B, C, D" and write "A, B, C, D" again, so I have two alphabets, and then I just shift one string of characters so that, for example, "A" goes to "D" and "B" goes to "E" and so forth; and whatever sticks outside, you just bring to the front. In this case, the letters "X, Y, Z" at the end get shifted to "A,B,C" respectively.

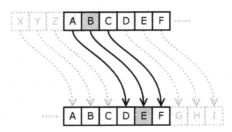

Substitution cipher with "shift 3" rule

Here the substitution rule is just "shift three"; and then when you get this cryptogram from me, you apply the rule "shift three" in the other direction to unscramble things and you're done: the message becomes clear.

Once again, prior to our secure communication we have to agree—we have to share a secret—and that secret, in this case, is the specific nature of the shift: how many characters we are going to shift the whole alphabet by. That clearly shows that secure communication is based on prior, common secrets that we share. It doesn't have to be anything particularly meaningful, of course: whether we shift the alphabet by three or five or ten or some random number—it doesn't really matter as long as it is known to you and me and to nobody else.

So these two examples demonstrate two very simple methods of secure communication: one is based on the permutation of characters and the other based on substitution. Mathematically speaking, those substitutions are also regarded as "permutations", by the way, but that's just a detail.

So now the question is, *How good are those ciphers?* Well, of course—if you take a substitution cipher, what is called a "monoalphabetic cipher" where there is a one-to-one substitution—for each character in my message there is another character—and I'm consistent, so each time I see "A", I'm going to use another letter or something else; it could be a graphical symbol, it doesn't even have to be a letter of the alphabet.

As you might expect, it turns out that, even though there are a lot of possible substitutions we can come up with for a simple alphabet, it's very easy to break monoalphabetic ciphers. And the way you do so is by way of a statistical test.

When we communicate in written form, all natural languages are not just a random collection of letters: we use the letter "E" much more frequently than we use the letter "Z", say. In most Indo-European languages, the letter "E" is the most common one, certainly that's the case in English.

HB: So, presumably, one could work backwards and say, for example, *"This symbol over here had a much greater frequency than other symbols, therefore, it is, in all likelihood, an 'E'—or at least it's very likely to be a letter typically used with great frequency."*

AE: Yes, that's right. In order to crack a message that was encrypted using monoalphabetic substitution, you'd look for the most frequent character and assume it's the letter "E". Then you'd assume that the second most frequent character is the letter "A". Furthermore, there are certain combinations of letters like "TH" that appear quite often and so on and so forth. So, there is a very simple, statistical analysis that will easily reveal the text.

Today, in fact, we don't even consider this as an encryption; one-to-one substitution is considered as switching from one alphabet to another, but it took people a while to figure this out.

Finding a systematic method of breaking monoalphabetic ciphers goes back to Al-Kindi, who was working in Baghdad in the 9th century. This was the period of great flourishing of Arabic culture when many ingenious discoveries in mathematics and other fields came out of Baghdad. If you were living in the 9th century, the place to be to work on any number of scientific and other intellectually important fields was definitely Baghdad.

Anyway, clearly this one-to-one substitution was not good enough. The next big step in the history of cryptography was a switch from monoalphabetic ciphers to polyalphabetic ciphers. People thought, *If one-to-one correspondence is easy to break, what about one-to-many?*

Perhaps one way to think about this is to imagine that, rather than using the Caesar cipher with one particular shift—say, three—we instead use a sequence of several designated shifts in our message. For example, for the first character, we use shift by three; for the second character, we use shift by seven, and for the third character, we use shift by ten. Then we have this "three, seven, ten" sequence that we keep repeating.

That's a little bit more complicated. Those kinds of ciphers were proposed in the 14th and 15th century in Europe during the Renaissance when everybody was plotting against everybody else, particularly in Italy.

HB: So there was a great need for it.

AE: That's right. There are a few names associated with the invention of the polyalphabetic cipher, but one of them was actually quite a famous architect and polymath who in many ways personified the spirit of the "Renaissance Man", Leon Battista Alberti. He came up with this idea of having two, concentric discs and on the edge of those discs, you would write the alphabet both on the inner one and the outer one and then, if you rotated one with respect to the other, you could see the substitution, which letter on the outer disc corresponds to a letter on the inner disc. So, that gives you a simple, mechanical device that allows you to remember the substitution and then you can use this to help with encrypting. So, you simply just rotate by three, seven, ten, whatever the sequence of letters might be that we agree on and that allows you to encrypt and decrypt.

A "cipher disc"

This ingenious idea looks very simple and almost unnecessary; it's just a tool to remember the substitutions and help you to try to decipher them on the other end. But then, with all those electro-mechanical devices in the 19th century, that Alberti disc evolved into a rotor machine with many discs with the corresponding letters.

HB: And my understanding is that the Enigma machine was really a large-scale and much more complicated device that incorporates all of these substitutional encodings?

AE: Yes.

HB: You have all these different rotors, and you can imagine them moving at different rates and so forth; but it seems to be the same basic principle at work.

AE: That's right: it implements a very, very complicated, polyalphabetic cipher where there are a number of different variables that you can control, but at the end of the day it's a polyalphabetic cipher.

It's not that easy to break polyalphabetic ciphers, especially if you use many different substitutions. We know how to do it in principle, but it requires a certain degree of statistical analysis. By the way, Charles Babbage was the first one who came up with the idea of how to break polyalphabetic ciphers in a systematic way.

HB: Which emphasizes the link between computing and cryptography right from the start, as it were.

AE: Yes. As you know, Babbage was doing all kinds of things, including cryptanalysis—how to break ciphers. If you push this polyalphabetic idea to the extreme—suppose I have a message and that has N characters—letters—in it, and I use N alphabets, N randomly selected alphabets for each letter, then we push the whole thing to the limit.

In other words, we have a polyalphabetic cipher where for each letter we use a different, randomly chosen, alphabet. This brings us to what was considered the ultimate cipher: the one-time pad.

That's one way to think about it, but probably an easier way to think about the one-time pad is to consider a message that you can write, say, in binary notation—of course expressing a message in binary notation is not an encryption, it's just changing the alphabet.

Then I compose a randomly selected series of 0s and 1s that I call "the cryptographic key", which is a very important concept in cryptography.

So the idea is that I write my message in binary notation and underneath it I put my cryptographic key, that random series of 0s and 1s I have created. And now I add the two together by simple binary addition—$0+0=0$, $0+1=1$, $1+0=1$ and $1+1=0$—producing an encrypted message.

Now my original message, even though it's expressed in binary notation as a sequence of 0s and 1s, still has the statistics of a natural language, which means that if you look at it closely, certain sequences of 0s and 1s are more likely to appear than others and it is thus able to be subjected to the sort of complex statistical analysis we were talking about a moment ago.

But when I chose a completely random sequence of 0s and 1s to add to it through the process of binary addition, that introduces a measure of randomness to the message from the randomness of the key. If you take something structured and add it to something random, then you get something random—it inherits the randomness of the random part.

So, the cryptogram, which is the sum of the message plus the key, is a truly random thing: you will not find any statistical regularities there.

1 0 0 1 0 0 1 0 0 1 1 1 0 1 0 1 0 0 0 0 1 0 0 0   Original message in binary notation (with statistical structure)

1 1 0 1 0 0 1 1 1 1 0 1 1 0 0 0 0 0 1 0 1 0 1 1   Cryptographic key (random)

0 1 0 0 0 0 0 1 1 0 1 0 1 1 0 1 0 0 1 0 0 0 1 1   Encrypted message (cryptogram) made by adding the cryptographic key to the original message (random)

Encrypting a message using a binary cryptographic key

HB: Which means that somebody on the outside, who's trying to break it is, in principle, not going to be able to break it because of the randomness that was incorporated in that process.

AE: That's right, yes.

HB: So, my sense of the situation is the following: if I've got two people who want to communicate something, as long as they both have access to this random key—and they're the *only* ones who have access to this random key—then it's impossible for somebody on the outside to be able to understand what's going on because this random key that they both have access to will produce sufficient amounts of randomness in the cryptogram, the encoded message, that any eavesdropper who might intercept this encoded message will not be able to make heads or tails of it.

AE: That's absolutely right. We know that, given enough computer power, we can do a statistical analysis on the original uncoded message that's expressed in binary notation to be able to figure out what's going on. In other words, the allocation of 1s and 0s is not random, it's what we call "meaningful". Meanwhile, I also produce my cryptographic key in binary notation that is *not* meaningful, because it's simply a random allocation of 1s and 0s.

In other words, what I have is a meaningful binary string and a random binary string. The meaningful string is the message I want to send to you: *Howard, let's meet in the pub around the corner for a beer.*

And to this message, this meaningful sequence of 0s and 1s, I add my cryptographic key, a randomly chosen sequence of 0s and 1s that has no statistical pattern whatsoever—it's just random noise.

And when I add the two together to produce my cryptogram, I also get something that is truly random. It's equally likely to be anything, it's as random as the key. I then send the encrypted message, this cryptogram to you.

Now you, and this is very important, have to have the key in order to make sense of things: you have to have exactly the same random sequence of characters that I used to add to my original message. We need to share this secret.

So once you receive the cryptogram, you can use the key to abstract the randomness that was introduced when I added the cryptographic key to my original message.

Now, assume that someone else is trying to eavesdrop. That guy, an eavesdropper, can write down the encrypted message, of course, but it simply looks like a random string of characters. Without knowing the key, you cannot do anything because there is no underlying statistical pattern to work on.

Now, there are a few assumptions here that we need to make explicit. In addition to being random, the key has to be as long as the message. Then it has to be secret: only you and I have this key, nobody else has access to it.

And lastly, we're only going to use it once, just for that particular message. Once we start recycling keys, then any potential eavesdropper who can collect enough cryptograms will start to see a statistical pattern emerge due to the fact that all of those cryptograms have used the same key: the situation will become like a polyalphabetic cipher—the situation becomes analogous to using a cryptographic key which is smaller than the original message, which enables some information to be extracted through statistical analysis.

In fact, there are remarkable cases where people tried to cut corners and got caught: apparently during the Cold War the Russians sometimes tried to recycle their keys, and the Americans were then actually able to decipher some of their messages.

Questions for Discussion:

1. How can we be certain if the cryptographic key that we are using is "truly random"?

2. To what extent does the development of cryptography serve as a good example of how human antagonisms have been an essential incentive to gain a deeper mathematical and conceptual understanding? Do you think that such developments have occurred even in the absence of wars and factional fighting?

*3. What do you think Howard is specifically alluding to when he refers to "another revealing example of how language can influence thought in some particular way"? Readers with a particular interest in this topic, often referred to as the Sapir-Whorf Hypothesis, are referred to Chapter 11 of **Speaking and Thinking** with UC San Diego psycholinguist Victor Ferreira, the Introduction to **The Psychology of Bilingualism** with York University psychologist Ellen Bialystok and Chapter 9 of **Sign Language Linguistics** with UC San Diego linguist Carol Padden.*

III. Public Key Cryptosystems

Harnessing the difficulty factor

AE: But now we come to this essential problem, you see, because, if we are supposed to throw away our keys after each secure communication, then how do we establish these keys in the first place? If you are somewhere in France and I am somewhere in the UK and we want to communicate securely, somehow we have to find a way to generate this shared randomness, which is known in the field as "the key distribution problem".

As I said a moment ago, if we start reusing the same key then it's not going to be secure in the long run, so the crucial thing—the Holy Grail of cryptography—is to find a way where we can establish these truly random sequences of 0s and 1s to be exclusively shared between you and me and applied to our one-time pads.

We need to find some ways of developing this "shared randomness"—to distributing this secret information to just you and me; the information itself is random, it doesn't mean anything in and of itself, but it's an essential tool to be able to successfully use these one-time pads. Shared secret randomness is an enormous resource for cryptography.

So how do we solve the key distribution problem?

There are two ways of doing this. The first one is known under the label of "public key cryptosystems," which is what we use today for banking and protecting electronic transactions.

The second way is through the relatively new area of quantum cryptography.

For public key cryptosystems, the idea is basically not to use one key, but to use two keys in a very clever way; and even though it's usually associated with a beautiful piece of mathematics, perhaps it's better to explain how this works using a mechanical analogy.

Suppose I have a safe box that has two key holes and two keys; one is for locking the safe box, which I will call "the public key", and the other one is for unlocking the safe box which I will call "a private key". Everyone is in possession of two keys, their personal public key and personal private key. The public key is used to lock the box, while the private key is used to open the box, and only the designated recipient of the box can open the box with his or her private key.

Now, imagine that I have this safe box in my possession. It is open, and I can distribute it to anyone I want. Suppose I decide to send it to you. I lock it with my public key and send it to you. Perhaps once I've locked it, I realize that I made a mistake and I want to open up the box and change the message. Well, it's too late, I can't open it, because only your private key can open the box, so I have to start over with an entirely new box.

OK, so I send it to you and now you can open it with your private key. Of course, in reality the safe box is a mathematical system, with the keys being numbers and the mechanics of opening up the box depends on the mathematical interaction of these numbers.

As I said, my public key is a number that is advertised in some common directory: everyone knows what it is and anyone can use my public key to encrypt messages for me and send them to me, but the mathematics of how the keys interact is what establishes the security of the system .The most common type of public key cryptosystems is called "RSA", which is an acronym for Rivest, Shamir and Adleman, the surnames of the people who came up with this idea.

In fact, the same cryptosystem was proposed a few years earlier not too far from here in Cheltenham by people working for the British Government Communications Headquarters, GCHQ, but they couldn't publish it at the time. It's now been established by historians, however, that they had the idea earlier, and it was rediscovered in academe.

Now, the obvious question to ask is: How secure are those systems? We have those two keys; and they are somehow mathematically linked to each other.

So suppose we are using RSA and I am sending a message to you. Meanwhile someone else, an eavesdropper, wants to intercept the box and open it, for which they will need your private key, or the mathematical information associated with using your private key. Can they do it? Can they break the system and figure out your private key from the information available? Is it possible?

And the answer is, *Yes it's possible, but it is extremely difficult.* There is now this whole beautiful area of computational complexity that comes into the game, where you divide mathematical problems into classes: some mathematical problems are easy and some are considered difficult.

So what does that mean?

Well, you define a problem as "easy" if you can find a systematic way of solving it—an algorithm—so that if you increase the size of your input then the execution time for a computer—how hard you have to work to get the answer—grows, but only as a polynomial function of the size of the input. On the other hand, if it starts growing exponentially, you then call it a "difficult" problem.

Take multiplication. You start multiplying one-digit numbers, then two-digit numbers, then three-digit numbers. And you can plot a graph of the problem under consideration and how long it takes to do it. On the horizontal axis, you would plot the number of digits that you are multiplying, and on the vertical axis you plot the execution time—the time it will take you or your computer to carry out the operation.

Of course, it takes you or your computer a little bit longer to do multiplication of ten-digit numbers than five-digit numbers, but the key question is: *How much longer?*

If the time required increases polynomially, it could be linear, it could be quadratic, but it's not more than polynomial—then you can say, *"Fine, we have enough resources, we can wait that long; or perhaps we have some other physical resource that we can add that is equivalent to time—maybe we can increase the memory, say, and handle this problem."*

But now let's take the reverse problem to multiplication, which is factoring. Suppose I give you a composite number and ask you to find its prime factors.

I'll tell you: "15" and wait for you to tell me the corresponding prime numbers that when multiplied together give 15. So for 15, you'll fairly quickly be able to answer: "3 and 5".

And if I give you a larger number, you'll clearly have to think longer. And if I do the same sort of thing as I was talking about before, if I keep increasing the size of the composite number and plot the number of digits on the horizontal axis and the time required to find the corresponding prime factors on the y-axis, this function is not going to be polynomial. It's going to be an exponential function.

It's not that *you* personally cannot do it, rather it's simply not known how to do it—*nobody* knows how to do it. We don't actually know if this is an inherent difficulty or just that nobody knows how to do it right now.

HB: So it's quite conceivable in principle that somebody could come along tomorrow and develop a new way of doing things, a new algorithm for how we can factor composite numbers into their corresponding primes in a much faster way, in polynomial time.

AE: Yes, that's right. It's absolutely conceivable that, one day, there will be this super-duper genius who comes and says, *"Look, I know how to do factoring in an efficient way."* That would be a breakthrough in computer science and it's not entirely impossible.

This factoring thing that we use for our safe, for public key cryptosystems is certainly something that many think is at least possible to do differently. You would probably have to do some kind of a survey among computer scientists to get a sense of how many of them really believe that we can find such an algorithm in principle, but factoring is not regarded as the most difficult problem of difficult problems. It's in this class where most believe that it might be possible—even more so, in fact, because we know that if we build a quantum computer, we will have an efficient algorithm for factoring.

HB: Right. Just a quick summary, then. My understanding is that all of our present cryptographic systems that we use for purchasing by credit card and so forth, are fundamentally based on this idea that identifying the prime factors of a very, very large composite number is a complex process that necessarily takes a large amount of time notwithstanding how fast our best computers are. In particular, it takes an exceptionally long time if we go to the trouble of making the original composite number incredibly large.

AE: It could take more than the age of the universe.

HB: Right, very long.

AE: Call it long, yes.

HB: And since doing this factorization is integral to cracking our cryptographic systems, we don't have to worry, because however fast our standard computers become—we'll get to quantum computers shortly—we can always simply increase the size of our composite numbers to stay well and truly ahead of them. So our security protocols are safe.

AE: Yes. I would just add for the record that everything we are discussing so far is from an academic perspective. In fact, the cryptosystems that we use today do not fail because someone managed to factor such and such number, but because people start using them in a stupid way or somehow act negligently.

HB: We have too many drinks at that pub, say.

AE: Exactly—and we started talking about confidential things. So, that is the problem, that is exactly the problem. The mathematical side is not much of an issue for practical cryptographers—their principal concerns involve bad implementations and negligence and not following certain procedures.

But keeping this discussion at the purely academic level, because this is what I'm interested in—I'm not interested in nitty-gritty engineering details and how we implement those things—what you said is absolutely correct: that the best cryptosystems we have today rely on the unproven assumption that certain mathematical operations are "difficult", in the way I just defined it.

Of course, as we said, one day someone may come along and say, "*Well, guys, it's not so difficult because here is the way to do it—it's possible.*" We cannot prove right now that it is impossible.

Moreover, there are now indications from quantum physics that quantum devices, quantum computers, can be programmed in such a way that they *can* factor large integers in an efficient way.

Questions for Discussion:

1. How many different "complexity classes" of mathematical problems do you think there are? Might there be an infinite number? Might it be impossible to know?

2. What are the details of the mathematical relationship between the sender's public key and the receiver's private key? (See, for example, the Wikipedia page for RSA.)

IV. Harnessing Interference

The power of quantum computers

HB: Let me ask you to talk more about what you mean by that. What is a quantum computer, exactly, and how can it be distinguished from a normal or "classical" computer?

AE: Well, first of all many people might naively think that, since technology is progressing so quickly, a quantum computer might be something that is a super-duper computer that is processing information so quickly that it can somehow turn a difficult problem into an easy problem.

But that's not the right way to think about it. No matter how much you increase the clock speed of your computer, you're not going to change your problem from difficult to easy, you're not going to change the complexity class of the problem, because it's all about scaling: if you rescale something that is polynomial by a linear factor it's still polynomial; and if you rescale something that is exponential, you'll never make it polynomial.

So progress in technology per se—building faster computers—doesn't mean that you can break cryptosystems, because it doesn't mean that you can take an exponentially difficult problem and make it easy just by rescaling things. It's very easy to find a slightly larger number that is going to be equally difficult for that faster computer.

So the first misconception that people have about quantum computers is that it's merely some form of technological progress per se—that quantum computers are just simply faster. That's not quite true.

The reason a quantum computer can take a difficult problem and make it into an easy problem is not because they are faster per se—that their clock speed is faster—but that they process information in a different way. There are certain instructions that do not make sense for a classical computer but make perfect sense for a quantum computer; and we can use this extra set of instructions to build different algorithms.

When you construct an algorithm, you have a certain number of operations that you can use. If you translate this into a circuit or a logical network, you know those are basic logical operations like logical AND, logical NOT, logical OR, and so forth.

There's a finite set of those; and you can construct any Boolean function, any operation on your computer is expressed in those very simple operations. So, you have this set of operations; and nobody knows how to find an algorithm within this set, how to build a circuit, that would give you efficient factoring.

However, with quantum computers, you have certain operations, like the square root of NOT, that are literally not possible to implement within classical physics. So all of a sudden, you have this extra operation that you can use, an additional operation that you can add to your "Lego set"—you have a few more blocks to use. And if you use them cleverly, you find that you can construct an efficient algorithm for factoring.

So quantum computers can do it simply because they bring an extra set of operations. Now, *why* do they bring an extra set of operations?

Because information is not a purely mathematical concept, it is something that always has a physical representation. When we talk about "operating" or "playing" with numbers, what this really means is that there's some physical process going on that is responsible for this "operation".

Computers are devices that you can kick, they are physical objects with physical processes going on inside them. And the questions of how information is represented by physical entities and how these physical devices process this information is extremely important, because if we discover new laws of physics we can then process this information in a different way using those laws of physics.

So the discovery of quantum physics, learning more about quantum phenomena like quantum interference or quantum entanglement has direct implications for information processing: there are going to be significant differences in information processing between classical and quantum devices because they operate under different physical laws and information processing is an inherently physical phenomenon. That's the underlying reason behind these results, like the quantum algorithms for factoring.

Now, perhaps I might make a small diversion that might be of interest to people.

HB: Go right ahead.

AE: I'm often asked—both by people inside and outside of the spy business—the following question: while it looks like factoring can be broken by quantum computers, are there other classical ciphers that *cannot* be broken by quantum computers?

Many government agencies—the NSA in the USA, the GCHQ in the UK, and so on—are seriously concerned about the future of public key cryptosystems, but we know that it's not only factoring: we can also use different mathematical operations to construct public key cryptosystems.

And for them, one of the big issues is this: let's suppose we do build a quantum computer. We're not completely sure if it can be done, or whether if it can it will be done in 10 years' time or 50 years' time, but it probably will be built at some point because at present there doesn't seem to be any physical law saying that it cannot be built.

So if you build a quantum device that can factor large integers and suddenly RSA is not good enough, then perhaps I can come up with some *other* mathematical problem that I can use for public cryptosystems.

In other words, *Can I come up with a mathematical approach that I can use for public key cryptosystems which is "quantum-safe"?*

That is a very, very good question that is of great interest to government agencies at the moment, which has naturally spawned an entire research area to try to address this question.

There are some other public key cryptosystems like elliptic curves or lattice-based cryptography or McAleese cryptosystems based on some error correction code, just to drop a few names, that people are currently investigating.

And it may well be that some of them will prove useful, in the sense that even if we have quantum computers, they will also have limitations and might not be able to solve those approaches. We're not sure.

At present we think that elliptic curves probably could be successfully addressed by a quantum computer, some aspects of lattice-based approaches perhaps might work, error-correcting-based cryptography—well, maybe.

We're not sure. This is an open problem. And I'm not quite sure about the details now, but I believe that the NSA came up with some guidance for cryptographers on how to design public key cryptosystems in the future where they have to take into account the possibility of building a quantum computer, so it should be quantum safe.

Which means RSA will probably be gone but it might be superseded by another public key cryptosystem such that we are a little bit more confident that quantum computers cannot break them.

But this is tricky, because we don't really fully understand the power of quantum computing and it will be extremely difficult to certify that this mathematical problem will *always* be difficult for quantum computers.

HB: Well, that's what I was going to ask. So, maybe I'm just being paranoid but perhaps paranoid is a good thing to be in this line of work.

Just as you were saying earlier that there is no way of knowing, a priori, that somebody won't come along tomorrow with some really clever way of factoring large numbers in such a way that it can be done in polynomial time, presumably there are also cases where just because people haven't yet developed a quantum algorithm to successfully attack a problem, that hardly means that it is impossible to do. All we can say is something like, "*According to the present state of our understanding, there doesn't seem to be any way to develop a quantum algorithm to solve this efficiently."*

AE: Absolutely. We are looking for problems that can be useful for cryptography, which tend to be those that are difficult to do one way and easy to do another way—not every difficult problem is necessarily useful for cryptography—and within this class of problems, as you said, it's extremely difficult to make a statement that such a class of problems will remain difficult for quantum computers.

Most computer scientists would probably say that some of those very difficult problems would also be difficult for quantum computers, but those kinds of problems may not be good for cryptography. In other words, those problems that we are talking about are typically somewhere in "the twilight zone" between difficult and easy—the whole area of computational complexity and classifying problems into "easy" or "difficult" is like a zoo: there are so many complexity classes.

To me that indicates that there's much to be learned in this field. My sense is that as the field progresses, we'll likely see reductions in the number of complexity classes, but right now it's like the early days of biology before we managed to classify species in a reasonable way: we're just naming many individual instances.

HB: We're doing complexity taxonomy right now.

AE: Absolutely, yes; I would say so. And the overarching goal is to be able to at least come out with a statement like, *"We strongly believe that this particular problem will be difficult for quantum computers, so it's good for public key cryptosystems—use it in the future."*

For the banking and military industries, it naturally takes some time to change the entire infrastructure in order to start, in a consistent way, using some other system. So, that's a big deal. That's where we are in trying to find good public key cryptosystems for the future.

Interestingly, what quantum mechanics takes away from this area, in terms of forcing us to look at different alternatives from our current framework that are quantum-safe, it also gives back through quantum cryptography.

As it happens, you can prove that quantum cryptography cannot be broken by quantum computers, but the problem we have with quantum cryptography, at least for now, is that it's not as flexible for cryptographic purposes as public key cryptosystems.

Public key cryptosystems can do all kinds of wonderful things for us. They can be used not only to protect point-to-point communication, but they're also good for authentication. RSA, for example, can be cleverly used for authentication by playing around with this idea of a safe box with two keys that I was talking about earlier.

Mathematically speaking, it's really very beautiful: you reverse the role of the keys, so that I lock the box with my private key and you can unlock it with your public key. So, for example, if I want to send a message to you, and it's important to ensure that it really comes from me, I can combine security with authentication by effectively putting the box with my message in it inside another box. And when you unlock the bigger box you can see that I'm the one who sent the message, and then you can use the combination of my public key and your private key to open the inner box containing the message as before.

The precise details don't really matter, but what's important to bear in mind is that public key cryptosystems are very useful because they can also do all kinds of other important things.

HB: Right. And can quantum cryptographic systems be used to do this sort of authentication as well? Is it a little bit clumsy? What's the situation there, exactly?

AE: Well, we know that some cryptographic protocols are just not going to work, while others are just clumsy, yes. So, today if you want to sell a quantum cryptographic system—and there are people in business selling quantum cryptography right now—the strength of quantum cryptography is point-to-point communication.

Suppose you have a class of computers, let's say a database, and want to synchronize it with another database, which you own in some other location. Now within any given location you're sure that the computers are well-protected, but in order to ensure that you can securely synchronize the data between them that's the sort of case where you might want to use quantum cryptography, because it's point-to-point: one line that goes from A to B. So quantum cryptography would be your choice if you really want to go to the extreme and make sure that it's provably secure.

HB: I'd like to return to a point you made earlier about the physical nature of information in the context of quantum theory. It's difficult to know how exactly to tackle all of these points because there's so much going on at once, but maybe one way to do so is to take a closer, high-level look at our understanding of the quantum world and how it differs from the classical world.

You spoke about how quantum computers are able to harness different mathematical operations in keeping with their physical distinctiveness, like the square root of NOT, but perhaps it's useful to turn to a more physical picture when discussing the distinction between the classical and quantum worlds, like the famous two-slit experiment or Schrödinger's Cat.

So without wading too far into the details, one thing that these sorts of thought experiments do is that they highlight that in the quantum world unlike the classical world, it is not right or appropriate to ascribe distinct properties to objects before you've made a measurement of them. Another way to say that, I think, is that it's a meaningful—indeed, accurate—thing to say that quantum objects exist in a superposition of different states.

So my hand-wavy way of thinking about all of this is that, in keeping with what you said earlier about the fundamentally physical nature of information, what a quantum computer is doing is to somehow harness this fundamentally different aspect of the quantum reality, capitalizing on this new possibility to meaningfully exist in a superposition of different states and extract genuine information from that. Is that a reasonable way to look at things in your view?

AE: It is. Of course, you know, there is no one way of thinking about quantum computing and quantum phenomena. Even in physics, we all agree on the equations when we write them down, but when it comes to explanations then we start using different metaphors and different ways of describing things.

So that's why you have this rather diverse field of different interpretations of quantum physics, but yes—what you said is certainly one way of thinking about it. When we teach quantum physics, one of the first things that we do is to talk about this double-slit experiment, which is relatively easy to demonstrate and relatively easy to describe, but not so easy to explain the details.

You have a situation where you have the basic example of quantum interference where you have a particle that has an option to take one path or another path, through one slit or another slit and eventually ends up somewhere—in a photodetector, say, if it is a photon.

Then we ask the simple question, *Which path was taken? Was it slit one or slit two that it used to go from the source to the detector?* And we learn that thinking about the situation in terms of "either this one or that one" doesn't quite work. Somehow, you learn slowly, both paths are taken. It's not at all clear what "both paths are taken" actually means—it's somewhat mind-boggling—but that's the way it is.

So we then try to describe this using the language of quantum physics. So if you label the slits "0" and "1" you have a simple binary system and then you look for ways of describing this weird idea that the particle took both paths to get to the detector.

The binary value was not simply 0, because the path through slit 0 was not necessarily taken, but it's not 1 either because it was also not necessarily the case that only the path through slit 1 was taken, so you have to find a way to represent both simultaneously.

In other words, this phenomenon of quantum interference leads you to a situation where, if you label paths or states by, say, 0 and 1, you can sometimes have a physical entity somehow represent both states at the same time, which is this notion of a quantum superposition.

Quantum computers, then, really act like a huge quantum interference mechanism. Let's now not think about a particle that originated in one particular point and goes into a photodetector, but instead let's think about a complicated device in some, what we call, configuration space. It starts in some initial condition that includes the input to the device and the internal state of the device which is represented by a specific point in this configuration space.

Now, a classical computation takes you from this point into another point, then another point and so forth, so you have a well-defined trajectory that represents your algorithm, where the final output is represented by the final point in this path.

So, classical computation can be expressed as this computational path in this configuration space that takes you from your input to your output. It's predetermined by the program: you move along the path—click, click, click—from input to output.

Now the next thing is to think about a randomized algorithm where you may not choose just one computational path but you might take one or two or three or four different paths. We use these kinds of randomized algorithms frequently in things like so-called Monte Carlo calculations, and they can be quite powerful, quite effective. Now you're concerned about the probability of reaching some final point in this space that you're looking for and you're trying to maximize that probability of getting there.

Sometimes you get errors. It's often easy to tolerate errors—in fact, every single time you use your computer, there could be some incoming cosmic radiation that causes an error, but we can correct for that so it's not a problem.

But now when it comes to a quantum computer you think about those different computational paths in a similar way that you think about the different paths that a photon can take through the different slits: we take all those computational paths going from the initial point to the final point. And you can have many of them that interfere with each other—quantum interference is much more powerful than simply adding probabilities coming from the different paths.

Regular probability theory tells you that if something can happen in two different, mutually exclusive, ways—if something can happen with probability P1 and something else can happen with probability P2—then the probability that the whole thing happened is P1 + P2: I add probabilities. It's one of those axiom improbability theories. If you have mutually exclusive events, you add probabilities.

Now, it turns out that this axiom is one of the most important Kolmogorov axioms in probability theory, but, of course, axioms and mathematical constructions know nothing about reality—they are just some mathematical concepts. We don't even ask them to describe reality; we describe physics by learning about it in a different way.

So nature doesn't know about this axiom, and it turns out that it doesn't work this way: there *are* phenomena where something can happen in two mutually exclusive ways—like a particle going through one slit or the other slit—and the Kolmogorov axiom doesn't work because you try to add the probabilities together and it doesn't work. Sometimes you get more than you would expect from the classical addition of probabilities, sometimes you get less.

So the way physicists manage to solve this problem was to introduce the concept of probability amplitudes—a complex number from which you can derive the actual probabilities that correspond to the physical situation—in such a way that it accounts for this interference we were talking about.

Sometimes this interference enhances the probability of a given outcome (this is called "constructive interference") and sometimes it reduces the probability of some outcome (this is called "destructive interference"), where the probability can effectively go to zero in some situations, which is very unusual in regular, classical probability—if you have two positive numbers P1 and P2, adding them together will always give you something larger than zero, but in this case, negative interference will actually lead to the suppression of probabilities.

Now we come to the case of quantum computers. Just like before, you have an initial state and some final state; and there are all kinds of computational paths that this device can take from the initial state, the initial configuration, to the final one. But since this is a quantum device you now add probability *amplitudes* associated with each path, rather than probabilities.

And the art of constructing quantum algorithms is to make sure that you have constructive interference on the right answers—so the probability that you find the right outcome is enhanced—and the probability that you get the wrong one is suppressed by a negative integer. So you have destructive interference with what you don't want and constructive interference with what you do want—that's the game.

So this is one way of describing the action of quantum computers. And now, the question is, *Where is this "extra power" coming from?* The extra power is coming from the fact that it is really the case that each computational path that is possible was actually *taken* by the computer—like in the double-slit experiment—each slit was, somehow, utilized.

So somehow we utilize all of these different computational paths—but then, of course, people say, *"Well, at the end, you have to measure something. That's nice that you have this beautiful quantum interference, but at the end of the day you make a measurement and you destroy this interference,"* which is true, that's the nature of quantum measurements: to destroy this interference.

If you think about this as a vast amount of parallel processing occurring within one piece of hardware as your computer is going through many different paths, where, given a function, it can evaluate the function of all possible inputs. Each path can represent a slightly different number—so you can compute f_0, f_1, f_2 and so on.

So you start in a superposition of inputs and you get a wonderful superposition of outputs, but then you perform the measurement and you may lose it. It's true that quantum measurement, technically speaking, destroys quantum superposition; but it really depends on what kind of measurement it is.

In this case, the reason why quantum computers are so powerful is that we choose very carefully what kind of measurement we are going to perform at the end of the computation. Quantum computers are not necessarily good at telling us what all the possible values of the function are, but they can tell us something about the global property of the function.

For example, I might be interested in the periodicity of a given function. I know it's periodic, but I don't really know what this period is, I'm asking for one number, say it turns out to be 42, but I naturally don't know what it is ahead of time.

In classical computers, in order to find the answer, you have to start computing the values of the function. You compute f_0 then f_1 then f_2—and by the time you get to f_{42} you realize that you've got the same function as f_0, so that way you've discovered that the period is 42. But you usually don't know how long you have to go before you find it, so you have to do all these computations.

Meanwhile quantum computers do all of this functional evaluation, and the measurement destroys all this information, but it can still give you information about periodicity. So, we're in a situation where we find out this one number, 42, the period of the function, without learning what the specific values of f_0, f_1, f_2 were, which we didn't care about anyway. In this case the measurement destroys information we don't care about.

HB: Right. You're choosing a filter at some level?

AE: Yes. We say, "*Well, fine, you asked me about the period, you didn't ask me what the value of f_7 was, so I'm not going to tell you that.*" In a classical computer, in order to get the period, I'd have to go through all the possible values, so sure I can tell you what they were if you're interested later, but that is extremely inefficient because you have to do all those computations.

Another way to say this is that the quantum computer "knew about" the value of f_7—in the final analysis you might not get this number coming out, but it *was* computed. From an ontological point of view, if you will, it was computed.

Because of the particular act of measurement that you selected, that value may no longer be accessible to you, but that's another story.

And at the end of the day you got what you really wanted, which was something else: the period of the function.

And this, by the way, is exactly how we solve the factoring problem using quantum computers, because factoring can be translated into finding the periodicity of a certain function.

Given a number N, you look at a certain function and you know that this function is periodic but finding the period of this function is as difficult as factoring that number in the first place.

In other words you've translated a difficult problem about factoring into a difficult problem about periodicity, except that this difficult problem of periodicity, which reflects some *global* feature of this function, is actually a kind of problem that quantum computers like.

And this is exactly how quantum factoring was designed—finding the periodicity of a certain function. And while we do not get the particular values of the function (which we don't really care about), we *do* get the period; and once we have the period, we can, by a set of simple, classical operations, translate that into factoring the original number.

So this is basically how quantum computers work: we construct this huge, multi-particle, quantum interference, and we build this quantum interference from simple units—there are many quantum bits involved, they all go into superpositions, not only internally but also within each other which we call "quantum entanglement"—and we use those classical quantum logic gates that turn all those superpositions into other superpositions.

That's a detailed way of looking at it, but if you want to have a global picture, then the way to look at it is to say that this device performs a massive quantum interference of different computational paths, and these quantum algorithms are designed in such a way that constructive interference occurs for what you're looking to find.

HB: They have a specific end in mind.

AE: That's right. So, the right answer, by the explicit construction of the particular algorithm, comes where you get that positive interference. And that's not easy—the whole art of constructing quantum algorithms is not easy. But it can be very beautiful. So, that's how quantum computers work, that's how they happen to modify the complexity classes, taking some difficult problems and turning them into easy problems.

Questions for Discussion:

1. What do you think the original developers of quantum mechanics would have thought about quantum computers and quantum algorithms?

2. Do you find the idea that "information is inherently physical" a surprising one? Are you convinced by it? Are there some aspects of this claim that you find somewhat ambiguous?

3. To what extent do you think the notion of "beauty" in mathematics can be objectively justified?

V. Quantum Sociology

What is quantum information science, anyway?

HB: I'd like to return to explore this idea of the physicality of information in more detail, but before I do, I thought I'd return to the anecdote of the difference between physicists and mathematicians that I know you like to use, the one with the light bulbs that you started with a while ago when I so unceremoniously cut you off. Perhaps I can now ask you to relay that lovely little thought experiment.

AE: Sure. For a long time modern computer science was regarded as a branch of mathematics with all of its abstract concepts. I think Rolf Landauer from IBM was really the first person, in the '50s, who stressed the point, when he said, *"Mathematics, fine; but computers are not mathematical objects: they are real, physical devices—you build them out of stuff, there is physics involved. Therefore, what you can do with information might actually depend on physics."*

So, a good way to illustrate that, once we learn a little bit more about physics we can do more in this area, is to talk about this puzzle which I like very much. I like it for two reasons. It's a good illustration of what I was just saying, that the more you learn about physics the more you can do in this area, but it's also a good way to reveal what you were talking about earlier when you were describing that there were two types of scientists—those who are more oriented towards the world of abstraction and those who are much more driven by intuition.

In this case, you can tell whether someone is going to be a good experimentalist or a good logician. At any rate, here is the puzzle.

There are two doors leading to two different rooms. In one room—call it Room A—you have three switches, which are all initially in the off position. In the other room—Room B—there are three light bulbs. You know that each switch in light A is connected to one light bulb in room B, but you don't know which.

Your task is to figure out which switch is connected to which light bulb. You're allowed to enter each room once and do whatever you want when you're there, but once you leave that room you can't reenter it. And that's it.

Now many people would say, "*It's not possible.*" In fact, most logicians would tell you, "*I can **prove** to you that it's impossible,*" and they would start explaining to you how there isn't enough information for them to possibly solve the puzzle.

And in a certain sense they are right, if you only look at things mathematically there *isn't* enough information to solve the puzzle. But then if you think to yourself, *Hang on, we're treating this puzzle in a very abstract way—we've taken physics away from the picture. But there are no mathematical light bulbs in the world—light bulbs are real objects—so let's bring physics back into our picture.*

And when you do that, you find that there *is* something you can do—the puzzle *can* be solved. Go into Room A and turn two switches on and then, after 10 minutes or so, turn one of the switches off. Now go into Room B. And the answer is clear: the light bulb that is shining is connected to the one whose switch was left on, while the one that is off but warm from having been on for 10 minutes corresponds to the one that was switched on and then off, and the one that is dark and cold corresponds to the switch you never touched.

So now you've been able to solve the puzzle by taking heat dissipation explicitly into account—it is extra knowledge, extra information, that you've used to figure things out that came from recognizing that these light bulbs are actually real, physical objects. And without that knowledge you simply can't solve the puzzle.

By analogy, we can look at our relatively recent discoveries of the quantum world as extra knowledge, extra information, that we can use to solve problems that might not have been possible before when we only had our classical understanding of nature to rely upon.

HB: Right. That's a great puzzle. I'm glad we waited—I think it had even more impact this way.

I'd like to probe you a bit further on this notion of information being physical. An awful lot has been written about this over the years—some of it very deep, I think, and some of it verging on the nonsensical.

I suspect that some people might well have been talking about it before the dawn of quantum mechanics, but they've certainly been doing so a lot more vigorously since. On the whole, frankly, it's not entirely clear to me exactly what it means.

It's fairly clear, given your example as guidance, that we have to take the physical world into account if we really want to understand things as best we can. Feynman famously pointed this out in the early '80s, when he argued that the appropriate way to simulate quantum systems was through quantum computers, which is a related idea but a bit different, I feel.

My interpretation of what you mean by "information is physical" is a stronger sort of statement, but perhaps it's not. It seems to me that—to use a heavy word—you're invoking some ontological aspect of information saying that the very concept of information is necessarily tied to the physical world.

I mean, clearly we're all physical beings and so if we're talking about the information associated with us and the physical world we have to take the laws of physics into account, but I'm not sure that is a different statement than simply, *We should do everything we possible can to extract as much information from our problem space as we can.*

Perhaps I'm not being particularly clear, so let me try to say something more explicit. Maybe, to use your example, the logician who tried to convince us that the light bulb puzzle was insoluble would simply say, "*Well, now that I think about it, the way that I first formulated the problem was wrong. I failed to take into account that there was another degree of freedom—the 'heat degree of freedom'— and by overlooking that essential degree of freedom I didn't build my configuration space correctly and thus couldn't solve the problem. So it's just an oversight on my part, a failure to appreciate how to best assess the information, not so much that we need to invoke some meta-physical notion of information."* Do you see what I mean?

AE: Well, they're both components. You could say to your computer scientist colleague, *"Well, look, how could I take into consideration this 'extra degree of freedom', as you call it, if I don't know about it because it hasn't been discovered yet?"*

I have to specify the problem within a given set of knowledge using the language and using the phrases that are compatible within that state of knowledge. Even now, there may well be additional degrees of freedom that we are not aware of, things we don't yet know.

What I'm saying is that as soon as we discover that they do exist, then we can incorporate them and make them even a part of our set of mathematical tools, just as there are certain logical operations that quantum mechanics can support that cannot be implemented by purely classical devices. And that leads directly to the question: *What is the status, the logical status, of a possible operation?*

Let's take another simple example—and this is a great conversation to have with someone who is a proper logician, a classical logician.

You say, *"I want to define a new logical operation in the following way: I have a logic gate operating on a single bit, and arrange them so that they act independently on the same bit: I take the output from the first gate and then act on that to produce the final output."*

In other words, I have two consecutive gates—one gate after another—and the combination of the two gates together will implement the operation logical NOT. That's why I call each of them "the square root of NOT", because when I combine them one after the other I get logical NOT.

So now you ask this classical logician, *"Is it possible to have such a logical operation so that, if I have two consecutive actions of this logic gate, I get logical NOT?"*

And the logician would reply, *"No way. I can tell you all the possible operations that you can have on a single bit. If you take a single bit, you can only basically do four different things: it could be*

1 at the output, it can be 0 at the output, it could be logical NOT, or it could be identity. There are four ways of dealing with one bit and none of them has this property that if you concatenate two operations it will give you logical NOT."

And he would then say that he could *prove* to you that such a thing does not exist. But then, if you take a simple device like a beam splitter and you label the parts 0 and 1, you can show that there are two physical devices that can be placed in sequence once after the other that can implement logical NOT.

So since there *is* a physical representation of this particular logical operation, as a logician, you would have to add it to the repertoire of valid, logical operations. And along the way you would probably have to change the language you used to describe other things in the past—it's simply the case that reality must kick in here.

HB: And inform your logical awareness, as it were: increase your degrees of freedom, as I described it earlier.

AE: Absolutely. Now, when it comes to explaining how it works exactly, that's another story. Of course it's not a classical output because, indeed, such an output doesn't exist within the limits of classical logic.

So you have to think that through carefully and recognize that what we are dealing with here is a superposition of 0 and 1 and so forth. And how you explain that, exactly—which interpretation of quantum physics you use to explain such a thing—is another story as well.

As you mentioned earlier, when it comes to those interpretations we might use a different kind of hand-waving argument to explain how things work, but we all agree on the equations. And people in the scientific community can become so passionate about those interpretations that they can sometimes be a bit aggressive.

HB: I'll say.

AE: In fact, I didn't realize at first that this is also a perception of people in mathematics and logic. I remember once having a dinner at some conference—I think it was in Erice in Italy, a wonderful location—and we were having this conference dinner with a mix of mathematicians, logicians and physicists.

I was sitting next to a colleague and friend of mine, Bob Solovay, a logician with this very deep and very characteristic voice. At some point he turned to me and said, *"Artur, can I ask you a very personal question?"*

He had wanted to ask me this quietly, but as soon as he spoke to me everyone suddenly went silent. And then he continued: *"What's your favourite interpretation of quantum physics?"*

So that just tells you what it's like—it's almost like asking someone's religion: *What do you really believe in?*

HB: This brings up another interesting point. You're involved in something that you refer to as "quantum information science". People throw different words around to refer to this—"quantum information theory", "quantum computing" and so forth—but the point is that in my experience there is a real spectrum of what this is all about, what you're really doing.

There are those who think that this is "real physics" and what we really should be doing, experimentally matching theory and experiment. There are others who disdainfully refer to the whole venture as little more than glorified engineering—important engineering perhaps, but engineering nonetheless. There are some who are particularly excited at the prospect that, by looking at different experimental avenues for quantum physics and rigorously probing them, we will shed vital light on our understanding of quantum mechanics and quantum theory. Lastly, there are people who maintain that quantum information science will genuinely inform our understanding of what the right interpretation or understanding of quantum theory is and finally help us comprehend what it's really all about. Where do you fall in that whole spectrum?

AE: I think there's a wide range of perspectives to consider, and it depends which particular group of people you are talking to. As you know quite well—more than me, I suppose—different communities will have different perspectives on quantum information science.

HB: I'm not sure about more than you, but I've had my share of opportunities to get a lay of the land.

AE: You've certainly had your share; so you know that this really depends on what you think is important in physics—it's a weird mixture of social behaviour and assessing value: *"How do I judge whether this given area is important and what kind of importance is it?"*

So let's talk about quantum information science. In my view it's an excellent topic because it has a little bit of everything. Certainly the task of building a quantum computer or specific quantum cryptographic system is very challenging. In some sense it *is* glorified engineering, but that's hardly the whole story, because strictly speaking that would mean that you know, more or less, what to build.

Sure, physicists can tell the engineer what to build, but they may not tell you how exactly that thing works. In the classical case once you know how to build things and what to build, you also more or less know how things work. But that's not the case here: it requires a little bit of a deeper explanation.

So there is certainly a little bit of everything. You have good engineering, but it's challenging. You have good experimental physics too. Unlike something like string theory, this is an area where you can do lots of exciting experiments—you can falsify certain conjectures and at the same time often find yourself presented with interesting and unexpected emergent phenomena that require new explanations and deeper understanding. The field is full of surprises on the experimental side, and there are certain questions that you'll be able to answer by running an experiment rather than merely simulating it or making a prediction using conventional, classical devices.

Of course there's the question of intellectual snobbery. You might ask, *Does it have something like the "deep purpose" and "meaning of it all" associated with something like cosmology?* Well, as you know, intellectual snobbery is always a factor for some people no matter what the task is, but I think the important question is, *What genuine insights does this give you into physics?*

And personally, I think it offers lots of very interesting insights. It's not that I'm determined to talk to you as a spokesperson trying to sell the field, but it's simply a fact that the language that computer scientists developed was a different, very useful, way to understand the underlying physics, enabling us to express and reinterpret many physical phenomena in terms of networks, logical operations, circuits, and so forth.

And then it's the case that the area of the foundations of quantum physics, which was always somehow neglected and was considered by many hardcore physicists as something to be pushed aside and swept under the rug—which was also your experience, right?— suddenly that became important to this field as people began leveraging the knowledge that was developed by the foundations of physics community and incorporating those ideas into quantum information science, connecting them with the real world.

So suddenly, even some hardcore physicist can now see that Bell's Theorem can actually be used for something—there's an engineer who can actually build some device directly harnessing its insights that can be used for something—and now people's perspective starts to change as they begin to appreciate that this field of foundations has real, practical meaning.

Quantum information science, then, elevated—or even "rescued from the fringe" perhaps, to some extent—the area of foundations of quantum theory: it was brought into the mainstream of quantum physics.

And it also has to be recognized that the language that people developed in quantum information science has been successfully applied in a number of different contexts throughout physics, from condensed matter physics to the information in black holes.

So somehow, all of this happened quite unexpectedly from fairly humble beginnings. I remember that in the early days of quantum computing, it was sometimes referred to as "the Oxford disease".

Rolf Landauer, for example, was always very sceptical about quantum computing, and he was very annoyed that all those talented people would go to Oxford and come back infected with what he called, this "Oxford flu"—they would come back from Oxford having lost interest in working on reversible computers or more efficient computers and only wanted to work on quantum computers. Only at the end of his career at IBM did he change his mind, admitting that there was probably something to it after all, most likely due to some gentle persuasion from the likes of his colleague and friend Charles Bennett.

But my point is that in those early stages people went into this field simply because it was fun to work in and had some very interesting components to it. I liked it because it allowed me to look at some fundamental problems from a different perspective.

I couldn't care less, honestly, whether quantum cryptography would ever be implemented or not. What I was really interested in was trying to understand the process of computation better and the role that physics plays in it, to what extent the physicality of information is necessary to understand how to do information processing in a different way, to what extent we are limited by the mathematical constructs—why is it that mathematics and physics work so well together, why does mathematics possess this power to explain things in the physical world? Those types of issues.

So that was certainly my motivation then, and still largely is—I feel a bit like a salesman for quantum information science, but it is certainly true that nowadays the field has a very attractive mix of different areas: a bit of philosophy, a bit of engineering, experimental physics, theoretical physics, computer science, and lots of interesting mathematics coming from cryptography and so forth.

For people who come at things from the more fundamental side, building a quantum computer is not really an end in itself. It will probably be built, eventually, and will doubtless revolutionize many other things. So that's all fine and good, of course.

But sometimes, for me, I think that what would actually be more interesting would be to discover that a quantum computer *cannot* be built for some fundamental reason. Because if that were true, it would mean that we would have discovered something truly new and interesting about nature that we didn't know before.

If we would come up against a physical law that says, *This device cannot be built*, then that would be fantastic. Of course, we wouldn't be able to build a quantum computer in that case, but by understanding that physical law and incorporating it into our understanding we would somehow be able to use it for even more powerful computation. So there's no way to lose in this game.

Questions for Discussion:

1. In what ways is the claim "information is physical" related to the mysteriously strong importance of mathematics for understanding physics? (Those interested in this topic are directed to Eugene Wigner's famous 1960 article "The Unreasonable Effectiveness of Mathematics in the Natural Sciences")

2. Are you surprised to learn that notions of "intellectual snobbery" and other sociological factors can play such a strong role in a rigorously mathematical subject like physics? Should you be?

VI. Quantum Metaphysics
And its concrete effects

HB: So, a couple of comments and some follow-up questions. First of all, I don't think you should, in the slightest way, feel any embarrassment or hesitation at being a salesman for quantum information science. You're supposed to be a passionate advocate for what you find interesting and what you believe in.

I certainly appreciate the benefits of quantum information science in terms of making people reconsider or readdress, not just the *utility* of something like Bell's Theorem, but thereby naturally driving us towards a deeper level of understanding of what, exactly, is going on in the quantum world.

Which brings up the obvious question of, *Well, has that actually happened?* It's clearly true that quantum information science has given an enormous boost to the field of foundations of quantum theory and made the very act of talking about these issues, if not fashionable, at least suddenly acceptable.

But I can't help wondering if, when all is said and done, that's really made any difference. Has it advanced our understanding, holus-bolus, in terms of getting a deeper, conceptual awareness and understanding of quantum theory? Has it informed our knowledge in some particular way? That's a very high bar, of course, and it should be stressed that I'm not making any accusations here, but I'm just wondering how you would answer that question, as well your sense of whether or not anybody else is asking the question.

AE: Your question has so many subquestions to it—it's a very good one, of course, and one could talk a while about it. You're touching on both philosophical and utilitarian issues.

I think that quantum information science has helped bring quantum physics to the interest of the public at large and made quantum physics much more popular—somehow this combination of computation and quantum physics just sounds very sexy to many people.

It also brought to light some philosophical problems that physicists and philosophers were trying to resolve to the attention of the public. Earlier you spoke about this idea of a "glorified engineer". Well, the glorified engineer will have to think a little bit—I mean, at the purely instrumental level an engineer designs a system and that's it, but trying to go deeper and understand this system requires thinking through questions like, *What is the nature of this quantum measurement that we have?* and, *How come is it different from a classical concept?*

A classical measurement, a passive observation, is rather simple: we believe that objects have properties, those properties are associated with objects; and the measurement gives you some information on what that preexisting property is. It doesn't affect this property in any way; it's just a process during which you learn about the property. If my shirt is of such and such colour, this colour exists independently of whether you look at it or someone else looks at it—it has some kind of ontological existence.

The quantum measurement, however, is generally perceived somewhat differently, and then there's a whole history of dealing with this subject going back to the founding fathers of quantum physics.

Some people believe that the act of the quantum measurement is quite different and cannot be understood within the framework of discovering the preexisting property or preexisting value of something, but rather plays a more active role insofar as the act of measurement somehow makes the measuring device interact with the system that you're measuring, because it is only through this interaction that you learn something, and this process of interaction will affect what you see.

There are all kinds of interesting philosophical questions associated with all of this: *Are you a realist? Do you believe that there is an objective, independent reality? Are you more on the positivist side of things, refusing to even talk about "reality" as such, only caring about what can be seen, experienced and measured?*

As it happened, historically, when quantum physics was born, the major intellectual trend in Europe came from Vienna, and the "Vienna circle" had considerable influence on many thinkers.

So, it's not such a great surprise that, people like Niels Bohr and Heisenberg found themselves within this kind of intellectual framework. It had considerable impact on the way they thought about quantum physics, with the so-called Copenhagen interpretation of quantum physics being strongly influenced by this positivism coming from Vienna. I don't subscribe to this view, as you know, but most physicists probably still do, even though I think it's changing now.

HB: In my experience most physicists just say that they do so that they can quickly move on to other topics. But perhaps that's changed now.

AE: Yes, I think it has changed. In many ways the field made these discussions respectable because, somehow, it does matter sometimes. As long as it was just a philosophical speculation, those hardcore physicists would say things like, *"Who cares? If I can't do an experiment that would resolve things this way or that way, then it's nonsense and we shouldn't be talking about it."* That's a very positivistic way of thinking about those things, and I think it was not useful.

Even in the early stages, you could see that those who really asked good questions, like Einstein and Schrödinger, came from quite a different philosophical tradition.

Einstein was a realist who always had issues with quantum physics, and Schrödinger was the same—he was the first one who was puzzled by entanglement, I believe.

Needless to say, Niels Bohr and Werner Heisenberg became big figures in physics and did lots of fantastic work, but the real fundamental questions that became even more important later on were asked by Einstein and Schrödinger.

So I think that, somehow, your intellectual mindset really *does* affect the way you think. You can do physics at the level of equations, but if you don't add that extra power of intuition and explanation then you don't quite get to many new ideas and new concepts.

HB: The EPR paper was when, exactly? 1935 or something, right?

AE: 1935, yes.

HB: Which is interesting in itself, as you were just saying. I mean the conventional view for many years was that by the time Einstein went to the Institute for Advanced Study he wasn't doing anything interesting anymore, effectively wasting his time with his search for a Unified Field Theory and generally just being a figurehead for this new institute in Princeton, out of touch with mainstream physics and resting on his well-earned laurels for all of his previous accomplishments like the general theory of relativity.

Of course, I certainly don't want to diminish any of his other magnificent work, least of all general relativity, which I firmly believe is nothing less than one of the supreme intellectual accomplishments in the history of humanity, so there is that. But it's undeniable that this 1935 paper with Podolsky and Rosen had little less than a transformative impact on our understanding of the quantum world through the pioneering work of John Bell in the '60s followed by experiments by people like Alain Aspect in the early '80s.

And quantum information science played a huge role in making people aware of the importance of that, so much so that I would guess that the EPR paper is probably Einstein's most cited work these days.

AE: Yes, that's absolutely right. This is the situation: here is this grumpy Einstein who had made all these great discoveries and is now somewhere in Princeton, New Jersey. He doesn't quite like the way people try to picture quantum mechanics; he can see that it's a very effective tool, but he thinks that it's not the end of the story: there has to be a deeper explanation for what's going on that doesn't simply rely on statistical predictions. In his mind it's like statistical physics—a field he also made significant contributions to, of course—where we agree to use certain phenomenological models because we don't have enough knowledge.

So to Einstein, it was like that: this new quantum physics is a provisional construct, and sooner or later, we will come up with something better.

And in order to demonstrate his convictions that quantum theory wasn't the whole story, he came up with many great challenges to the theory. Some of them were answered by Bohr, but as you pointed out, trying to read Bohr and understand what he actually said is almost impossible.

EPR, the paper he wrote with Podolsky and Rosen is a beautiful paper. I came across it as a student and I really loved the way it was written: it was so clear. Very much unlike Niels Bohr, Einstein had this amazing way of presenting things in a very lucid and simple way.

Now, that paper was pretty much neglected by most physicists for a long period of time. It was only in the '60s that John Bell looked at it and tried to resolve the challenges that were presented there.

HB: And Bell was only doing this part-time, right? He was doing it as a hobby, effectively.

AE: That's right. He was supposed to be designing accelerators, not doing all this philosophical mumbo-jumbo, but he was genuinely interested in the foundations of quantum physics and he managed to take the ideas that were in the EPR paper and translate them into something that was actually testable in a lab, where certain questions about whether something exists in a conventional sense can be actually resolved in an experimental way.

So you might think, "*Wow, fantastic! What a deep and profound idea—there must have been zillions of physicists jumping at the chance to follow this up!*"

But that's not true at all. Nobody cared about this paper by John Bell. It was only another ten years or so later that a few people—John Clauser at Berkeley and a few other guys on the east coast—decided to set up an experiment to test the so-called Bell inequalities to see whether or not we could attribute an "element of reality", as Einstein called it, to certain physical properties.

What Einstein was basically saying was, "*Look, if you take quantum mechanics seriously, then I can show you that there is a problem with the definition of reality.*"

He defined this "element of reality" and showed that it may not exist; and these experiments that were designed to test that could be phrased in terms of those so-called "Bell inequalities", where you have a certain formula and you empirically determine whether it is more or less than some value.

And even at the very beginning there was a little bit of drama— even though not so many people were paying attention to this at the time—because you had one experiment on the west coast of the United States, which declared that the Bell inequalities were violated, so there seemed to be no element of reality, which was puzzling.

But then there was another experiment on the east coast that maintained that the Bell inequalities were *not* violated, so everything was as people thought it should be: you could straightforwardly attribute elements of reality to everything.

And then there was another experiment in Texas before Alain Aspect came along with his beautiful experiments in the early '80s in Orsay in Paris.

Alain Aspect is a brilliant experimentalist; and I have to say it was his experiment that really inspired me to look more carefully into those issues. Aspect's experiments conclusively showed that, yes indeed, those Bell inequalities *are* violated.

But even then, not so many people really seemed to care. Most people said, "*Well, so what? At the purely instrumental level we can still use quantum mechanics just like we did before, so what difference does it make, really?*"

There were a few people, of course, who thought about it, but on the whole not so many physicists did, and I find that extremely surprising. Here is such a deep and profound question about the nature of reality, and almost nobody is really reacting to this in the way they should.

For me, it had this huge "wow factor". All those guys—Bell, Aspect, and all the colleagues who did all those beautiful experiments before and after Aspect, are my heroes because they really changed our understanding about the world.

Unlike so many other experiments in physics where you *know* what's going to happen—where you design some experiment in order to "confirm" something we already believe in, to find a result that follows the theoretical prediction so that you can declare that "the experiment worked"—in this case you have a *real* experiment where you simply *don't know* what's going to happen, which is the point of a good experiment.

Here you have Alain Aspect sitting there in his lab, after those two experiments in the United States that produced these conflicting results, and he has no real idea what's going to happen with these Bell inequalities, but recognizing that whatever the result, it will change our basic knowledge about the nature of reality. It's fantastic.

I only had the chance talk to John Bell once and tell him about my idea of using Bell inequalities for the purpose of cryptography. And he was shocked. "*Are you telling me this is **useful**?*" he asked me, dumbfounded.

And I said, "*Well, I don't know how useful it might be eventually but this is what I'm working on. It's my PhD thesis.*"

HB: It was useful for you.

AE: That's right. I got my PhD out of it, so it was certainly useful for me.

And then it just took off. So I have my own contribution, I suppose, to making EPR more widely known because of my approach to quantum cryptography, but it should be recognized that quantum cryptography doesn't have to be based on EPR. I focused on this piece of physics because I happened to have found it fascinating, but prior to me my colleagues Charlie Bennett and Gilles Brassard used other physical phenomena to implement quantum cryptography.

I was simply driven by philosophical curiosity, saying to myself, "*Well, if something doesn't exist, there's no element of reality; and if information is physical and thus encoded in something, and you've encoded it in something that is, somehow, losing its element of reality, then there's nothing to eavesdrop, because it doesn't exist."*

So that was my thinking. And this demonstrates that quite often progress is not driven so much by rigorous mathematics, it's driven by these other crazy things, like "an element of reality".

In his EPR paper, Einstein defines this element of reality very clearly. He says, *"If in any way without disturbing the system, you can learn about the value of a physical quantity, then you attribute an element of physical reality to it."*

So the notion of "disturbance" was clearly specified there, and I thought to myself, *Well, this is exactly the definition of eavesdropping in cryptography.* To me, the definition of the "element of reality" in the EPR paper was reading like a definition of perfect eavesdropping.

And then I thought, *But wait a minute, now we know something about the non-existence of this element of reality thanks to Alain Aspect and John Bell, therefore if I exclude the existence of the element of reality, I exclude the existence of perfect eavesdropping.*

That was my independent way of thinking about designing a quantum system that would make perfect eavesdropping impossible—or, you could say, any attempt at eavesdropping detectable.

I was not aware at the time that Charlie and Gilles had their own ideas, but my work was straightforwardly motivated by those ideas. Which is interesting, I think. Here we have, effectively, this philosophical paper written by Einstein, Podolsky and Rosen—there's hardly any heavyweight physics there; it's only a couple of pages or so and is pure philosophy to most physicists. Nonetheless, if you start with those ideas you can really make some significant progress and do very interesting things with it, turning it into something extremely useful.

And as you pointed out, the whole issue of entanglement that appeared in that original EPR paper—even though I don't think they ever use the word—was eventually recognized as an enormous resource. This weird behaviour can be harnessed, tamed and exploited: we can put it into good use in cryptography, we can put it into good use in computers.

So all of a sudden, if you look at the citations of this EPR paper, after the '90s, it just increases exponentially.

HB: What's particularly fascinating for me is how you were philosophically motivated directly from EPR, directly from Einstein's paper. Good for you that he wrote so lucidly, by the way.

AE: Absolutely. I don't think I would have gotten anything from Niels Bohr's papers.

HB: Well, I'd have to question your thinking if you did.

We've been talking for some time now, and I really should stop, but I'm not going to because I'm having way too much fun. Too bad for you, but I've seen what remarkable energy you have so I'm going to take advantage of it.

So a few more questions. Several times during this conversation you've touched on aspects of what Eugene Wigner famously called "the unreasonable effectiveness of mathematics"—this notion that the world seems to be inherently mathematical to some surprising degree; and on the other hand you've also spoken at some length about the relationship between the physical world and information, the physical nature, or inherent physicality, of information.

Do you think that these things are linked in some way? They strike me as two different sides of the same coin, if you will. Is there a way that you can put together the mystery of why mathematics seems to be omnipresent in our understanding of physical law, and at the same time there's some sort of inherently physical aspect to the otherwise primarily mathematical concept of information?

AE: Well, of course I cannot claim that I have the answer to this question.

HB: Oh, come on. I'm not asking for "the answer". I'm asking for your perspective.

AE: Sure. So my perspective... Some time ago I wrote a paper together with David (Deutsch) and our colleague Rossella Lupacchini on exactly this issue. There are various ways to address Wigner's question of why mathematics is so amazingly effective, at least in physics.

The trivial type of response that many people would likely give, is something like, *"Well, we are physical entities embedded in the physical world; we see patterns in nature and abstract this by constructing mathematical theories based on our observations of the natural world, so it's no surprise that in the final analysis those mathematical constructs are so appropriate to describe what we see."*

That would be one way, perhaps the obvious way, of describing the situation.

Then, you may also ask, *"Well, do we really invent these things? Is there actually an ontological status to all these mathematical statements that we make? Is the world of mathematical concepts and ideas somewhat 'real'?"* And that leads us into one of the Popperian worlds, where those concepts are very real.

We may not perceive them through our senses like we perceive physical entities, but it could be that our mind has access to this Platonic world, and somehow, by doing mathematics, we are really operating in this world. But the way we do it is through some kind of a physical window, so that may explain, to some extent–

HB: Okay, I'm going to interject. I think somehow you felt that I was asking you to hold forth on all sorts of possible philosophical approaches, which wasn't my intention.

A large problem, you see, is that I'm often lousy at asking direct questions, so I'll redouble my efforts to be more direct. What I really want to know is what *you* believe.

There are issues of formalism vs. Platonism in mathematics. There are issues around the extent to which information is physical and what that really means, or at least what the implications of that are—some of which you've insightfully revealed in terms of your philosophical motivations that led to your pioneering work in quantum cryptography. There are questions related to the impact of your interpretation of quantum theory and earlier you spoke about your eventual adherence to the many-worlds view under the influence of David Deutsch and how things fell together for you there.

Here's my agenda. I'm trying to understand how Artur Ekert looks at the world and to what extent his scientific experiences have informed his philosophical worldview and vice versa.

More specifically, what do you really believe is going on in terms of information, in terms of mathematics, in terms of what's out there? Do you believe, say, that since information is physical, we gain access to this mathematical, Platonic world in that way? Or something completely different? You don't have to justify anything and I'm not looking for a complete answer; I'm just trying to understand how you approach things.

AE: OK, let me try to answer your question as honestly as I can. Even though I got those first ideas from the EPR paper where Einstein is questioning those elements of reality, even though I was motivated to draw conclusions from that work for cryptography, I have to say that I do not necessarily subscribe to the view that there is no element of reality there.

I think the problem is that Einstein defined this element of reality in one particular way. He never thought about a multiverse as a possible aspect of reality, because the Everett interpretation came later.

I think that if you agree that reality can be a little bit more complex—it's not just one single universe but it does have this multiverse component to it—then you can reconcile many things that Einstein would like to see by changing the definition of the element of reality so that it's not confined to one reality—you can have many. I believe, then, that Einstein can be, in a way, vindicated by this interpretation.

So even then it's usually the case that even in my public lectures I give a sort of instrumentalist overview of how you can use these philosophical concepts combined with Bell's Theorem to develop ideas about cryptography, this is not really the way I think.

Because, you see, in Bell's Theorem, there is this measurement at the end, and it's important there that you have one outcome of this measurement. Somehow the measurement is almost mathematical: it's not described by even the best physical theory. And then you'll see all those full contradictions.

We started talking about Bell's Theorem and how I was fascinated by it in the early stages of my career, but now I can say that, while I'm still fascinated by it, I don't think Bell's Theorem really represents what's going on.

The best physical explanation has to really quantize those devices, quantize everything, because there is no statement about where you should stop. So, if you quantize your apparatus, you start quantizing yourself, you put yourself into those superpositions—it's perfectly sensible—and then, in this broader view, there is not just one single outcome, but all possible outcomes can happen—they just happen in different Everett universes.

So I think that, even though I consider this notion of quantum key distribution based on Bell's theorem as one of my—probably my best—achievements in whatever I've done so far in my scientific career, somehow philosophically, I am throwing this away now, effectively saying that, "*Fine, it works, but this is not really what's going on.*"

Probably what is going on in order to explain this notion of security is that, if we take this view that there is a multiverse and that we exist in the multiverse, secrecy has to be redefined in terms of us not being correlated with information across different multiverses. In other words, I think there is probably a deeper explanation to this.

So that's part of the answer. I'm actually going against my previous narrative now and saying, "*Fine, it worked for me. I was motivated by this whole puzzle about the element of reality and used it to design a new cryptosystem. But in fact, philosophically speaking, I think there **is** an objective reality—it's just more complicated and should be expressed in terms of the multiverse, which would reconcile many things that Einstein would probably like to see.*" That's my current position.

And in terms of mathematical perception, I think you summarized it properly: I somehow do believe that we have access to this world of mathematical ideas through physics, hence the amazing effectiveness of those mathematical things.

And I do believe that we *discover* mathematics rather than invent it. I do believe that, when we discover a new theorem, we discover it, we don't just invent it—it's *there* and it's a shared experience that is accessible to all of us. It's not as tangible as this material world, but it's there nonetheless and we can talk about it.

The two of us can have a beautiful discussion over a more tangible thing like a glass of wine and discuss all those mathematical things and we would talk about it as if we could see it, sense it, touch it, explore the symmetries, properties, everything.

So it was not invented by me or by anyone else because in that case it would take forever to explain what it is. It's like finding something material and saying, *"Here, check this out,"* and everyone can look and see the beauty of it.

Questions for Discussion:

1. Would physics make more progress if philosophy was required training for all physics students? Less progress?

2. What do you suppose Artur means, exactly, when he refers to "Popperian worlds" during this chapter? How do "Popperian worlds" differ from the "Platonic world"?

3. To what extent is it possible to utilize a philosophical interpretation to make progress in physics without actually believing in that interpretation?

VII. The Joy of Questioning

And the merits of being rebellious

HB: I'd very much like to get to that (real) glass of wine, but before we do, a couple more questions. I'll start with the standard one I ask people of a scientific persuasion: if I were God and I could answer any three questions that you might have about the world, what would they be?

AE: Well, everyone would probably like to know how this universe really works. My hunch is that it's going to be something so simple that it would be really good to know what it is. So I guess my first would be something along the lines of, *What is the ultimate explanation for everything?* I know, I know, it's 42.

HB: That's the answer. It's just not clear what it's the answer to.

AE: That's right. Well, I suppose I would like to know what the ultimate physical theory is, and if there even is one. I hope there is one; I believe there is one, and I also believe it has to be very simple.

The progress of knowledge is not to make things more complicated—better explanations are more concise, so with just a few assumptions we can just explain more. So I would like to know what that ultimate physical theory is. That would be great.

Another thing that I would like to know, since you brought God into this picture, is related to the fact that, as a practicing scientist, I tend to ignore the non-material world. I believe that it probably does not exist, aside from things like the world of mathematical ideas and so forth.

But sometimes I realize that what drives me is this irrational magic of the beauty of things. As a child, you grow up and the world is full of mysteries and you want to discover them and they look like magic to you. Then, as we grow older, these things get more and more rational explanations, and somehow this magic fades away.

So what I would like to ask is, *Is there magic out there that we cannot perceive? Is there, perhaps, a world that is never going to be given to us?*

Another thing that I would like to know is, I still find it amazing that here we are, Howard, the product of these random forces of nature; and 70,000 years after cognitive evolution, we are here talking about and discussing this world.

How did that happen? At which point? Not only how did life start in terms of simple reproduction and so forth, but more specifically I would like to know, *How did Homo sapiens acquire this ability to understand the world in the way we are doing: asking for explanations? What is required for that to occur, exactly? What kind of components are required?*

And the last thing I would probably ask, would be the question about free will: *Do we have one or not? Are we responsible for our actions or not?*

This question has been asked by theologians and philosophers and it's not even clear what exactly it means, whether it can be reduced to randomness or some purely computational inability to be able to predict your actions in the future.

You may be able to tell me, or you may be able to see what I'm going to do, but for me it might be a different story entirely. In fact, it could actually be a physical system that is not necessarily me.

Suppose you design a deterministic system that has a certain degree of complexity—say a simple computer or even a mobile phone. If you take a mobile phone, there is a processor there and the processor has to assign memory to a certain computational task. Suppose you ask this device how much memory it is going to assign to a particular task a minute from now.

And on the outside you might know the answer, because it's a program that you designed, but for the device itself it's a kind of self-referential question: it has to simulate itself, and within this simulation it has to simulate itself again, and again and again, so there is a kind of cascade into self-referential questions, an infinite loop. And it's known that, in order to make approximate predictions about what is going to happen at time t is going to take at least t^2 amount of time—that is, considerably longer to make that prediction than the time required.

So, it could be that we can explain the notion of free will in terms of purely deterministic systems. It's possible. But that's exactly what I would like to know: *What is free will, exactly? Where is it coming from? How should it be understood?*

I wouldn't be asking questions about "the meaning of it all", because I don't know what "the meaning of it all" is—whether there is a meaning and a final destination, why things have happened the way they did and why we keep going.

But sometimes, I just feel that it's a weird fluke of nature that here we are, entities that, somehow, comprehend these things around us—at least we think so. We think that we have something interesting to say about this world; and we share those experiences and we construct wonderful devices and something interesting is happening.

We find ourselves intoxicated by the beauty of this world when we come up with something beautiful. Yet, somehow, we don't find, within this system, the meaning, the purpose—we have to go *outside* and talk about "religion" or something; and even then, the answer is not entirely clear: living forever, going to heaven or going to somewhere else, what does it really mean? Why would that be the meaning? Why should death have anything to do with meaning? And then there are related issues to cosmology, of course.

I don't know. There's a lot to ask. If I really was in this situation—if I really had access to God or the ultimate oracle, one thing I'm certain about is that I'd like to have a much longer conversation and ask many more questions.

HB: That sounds completely reasonable. You should know that of all the people I've asked, I think you've taken this question the most seriously. For once I feel genuinely sorry that I'm not God. It's almost like I've let you down somehow.

Normally, at this point, I ask people if there's anything that they'd like to say that we haven't had a chance to talk about. My sense is that we're just getting started, but we probably should start to curtail things. Instead, something else: Is there anything specific that you feel that we should have talked about and didn't or that you'd like to talk about a little more?

AE: No, I think that we covered it all pretty well. But earlier you mentioned that you might also want to talk about specific suggestions for teachers and students.

HB: That's right—thanks. I'd forgotten about that. What advice would you give students and/or teachers?

AE: Well, on the student side, I would just urge people to challenge authority; that's the best way to go. If you have a natural curiosity and you want to learn how things are, don't trust any authorities, because if you do, quite often you'll simply stop there.

As you know, I've been working in Asia for a long period of time and I noticed that there's an interesting cultural difference between Asian students and those in Europe and the US. Basically, you see that students coming from the West have fewer problems with authorities.

It's still the case that in a Confucianist Asian culture there is this notion of the teacher who has the ultimate say, and you have to respect your teacher. Of course, that's fine at the human level—everyone likes to be respected—but offering too much respect for something that a teacher says is sometimes not so productive in terms of promoting creative behaviour.

So, in this sense, modern science, where you question just about everything, is a purely western construct—it's a 17th-century European thing. It doesn't, of course, mean there hasn't been fantastic progress in the East as well, but modern science comes with a natural component of questioning authority and not taking anything people say at face value—"*nullius in verba*" is the motto of the Royal Society: take nobody's word for it. It's all about, "question, question, experiment".

For students, then, I would urge them to regularly question all authorities. Don't trust any authorities. Have your own view on things.

HB: Before we move to teachers, has that changed at all? You've been in Singapore for some time now and I imagine that you've seen considerable evolution.

AE: Yes, certainly. Science is more and more global. For example, most Singaporeans would travel and stay in North America and Europe for a while. So it has been changing a lot.

HB: But I don't mean so much at the postdoc or graduate student, level or even senior undergraduate levels, but, say, high-school students. Is there, generally speaking, an increased willingness to question authority in the younger students coming through?

AE: Yes, I think so. Even in those neo-Confucianist communities where the ruler is there to be respected and you should obey the ruler (of course the ruler also has responsibilities as well)—Singapore being one of them—you can see that this natural tendency to question is increasing among the younger generation. And that is good, I think. That's very good.

For teachers, on the other hand, I would just urge them to respect a certain degree of non-conformity on the student's side and to be patient. I have to say this because when I was growing up, I was not such a well-behaved child. I was pretty much a hooligan, in fact.

HB: You were a hooligan?

AE: Yes.

HB: I don't have you down as a hooligan.

AE: Well, you didn't see me when I was a kid. I had all kinds of problems with being rather arrogant and doing all kinds of things which put me on the border of respectable behaviour—perhaps even close to having a criminal record, in fact.

It was nothing really unpleasant, but I was not well-behaved, I would say, and I certainly couldn't stand all those pretentious and stuffy teachers. Maybe I've always had this rebellious nature, but I've quite often thought that a less regimented system would provide a better education than what I saw.

People have all kinds of views and opinions on this, and I recognize that different things might work for some people and not for others. I guess what I'm saying is that there is no "gold standard" for education and it is important to be prepared to try all sorts of different paths.

Personally, I'm very much in favour of education which is not regimented and allows you to explore your persona as you develop— that would be the kind of education that I would have loved to have.

So in summary to the students I would simply say, *"Challenge authorities"*—that's the only message I would give.

I wouldn't know what to say to teachers and how to design a good educational system, except that I'm convinced that there should be diversity, because there is no one size that fits all.

HB: You do recognize, of course, that by sitting here as a leader in your field and an authority figure explicitly invoking the motto "challenge authority", you're being somewhat paradoxical, because if they do listen to you...

AE: Yes, there is a beautiful logical inconsistency in what I'm saying. Let me start again. *"Please ignore every, single thing I said here in this discussion with Howard and just challenge every, single word. And don't write any emails to me because I will probably not reply, but think about it yourself. I was just providing food for thought, nothing more than that."*

HB: Perfect. Thanks very much, Artur. That was most enjoyable. I'm ready for that glass of wine now.

AE: You're very welcome, Howard. Me too.

Questions for Discussion:

1. Is rebelliousness an important ingredient for scientific success? Do some academic disciplines require more rebelliousness than others, on the whole? For an additional perspective on this issue, see Chapter 2 of **Pushing the Boundaries**, *the Ideas Roadshow conversation with Freeman Dyson.*

2. Are you surprised that a highly accomplished physicist like Artur would wonder about whether or not there is some sort of magic out there that we can't perceive? How common do you think such sentiments are amongst professional scientists? How might this be more a reflection of the fact that Howard and Artur clearly know each other than anything else?

The Physics of Banjos

A conversation with David Politzer

Introduction

Dancing To His Own Tune

It took a bit of time to get David Politzer to agree to sit down and talk with me. A co-recipient of the 2004 Nobel Prize in Physics, he was all too familiar with people approaching him out of the sheer excitement at being able to tell their friends that they had spent some time chatting with a Nobel Laureate.

That wasn't my problem. By a curious twist of fate, over the years I had found myself spending enough time in the company of Nobel Laureates to know that it could often be a highly overrated experience. Indeed, other than a few notable exceptions—such as the almost overwhelmingly genial Tony Leggett—knowing that someone had a Nobel Prize invariably made me look searchingly towards the nearest exit.

Still, it was David's Nobel Prize that first brought him to my attention, as it were. Years ago an astute colleague urged me to read *"The Dilemma of Attribution,"* David's clever, thoughtful, and humble Nobel Lecture that detailed the communal nature of frontline scientific inquiry, using his own "Nobel-winning" work on asymptotic freedom to explicitly demonstrate how science builds higher and higher edifices of understanding by compiling insights by different researchers, one upon the other.

It was the sort of thing that leaves a deep impression. So years later, when I began thinking about people to talk to for Ideas Roadshow, David was never far down the list, despite the fact that I'd never met him and just about the only things I knew about him were that he was a co-discoverer of asymptotic freedom, had worked under the legendary Sidney Coleman, and had penned a uniquely incisive Nobel Lecture.

When the time came to take a closer look, however, I was astounded to discover that his current line of research was focused around something called "banjo physics". A euphemism, perhaps? No. The lion's share of David's scientific efforts these days, it seemed, were focused on what, exactly, accounts for the banjo's unique twang.

Well, this was interesting. And a few more moments of googling revealed that, contrary to my first impression (no banjo aficionado, I, it must be admitted), there seemed to be a surprising amount of captivating science to learn about here.

Immediately, and not for the first time, I was reminded of one of the many fitting quotes by the legendary American physicist Richard Feynman: *Everything is interesting if you go into it deeply enough.*

The comparisons didn't end there. Feynman would routinely tackle a wide variety of scientific problems, spurred on by his insatiable, childlike quest to simply "figure things out", flagrantly indifferent to— indeed, at times, often actively hostile to—whether his colleagues considered the topic sufficiently "fundamental" enough. He was also, famously, a music lover, and a long-time faculty member at the California Institute of Technology. And along the way, he too, of course, had picked up a Nobel Prize (in 1965) for his seminal contributions to particle physics.

Of course David is not Feynman—nobody is. But for me, at least, talking to this excitable New York-born, Nobel Laureate in his Pasadena office a few doors down from where his erstwhile colleague revolutionized our understanding in so many colourful and unique ways, the spirit of Feynman was always with us.

No surprise, then, that we began our conversation with David enthusiastically relaying to me some heartfelt reminiscences of what life was like working with Feynman. It was a joyous celebration of everything I loved about physics and the human spirit.

And then we turned to the banjo, where there was no shortage of intriguing and unexpected revelations to be had, from how an open-backed banjo is like an ocarina, to the different effects of string-stretching, to the peculiarities of coupled, damped oscillators to how properly understanding sound naturally involves taking into account how our brains work.

And through it all David's unrelenting, unfiltered, curiosity shone brightly through. Over and over again, whether he spoke about the challenges of mathematically characterizing banjo sound or ruminating over the notion of information loss in black holes, he'd stop for a moment to forthrightly declare, *"I just want to know what's happening there. What's **really** going on?"*

At one point, when discussing some of his work in condensed-matter physics, he explicitly detailed his approach:

> *"There were questions there that I thought we should be able to answer. It wasn't hugely speculative. There was a way to carve out some very straightforward, Politzer-like, question that would have an answer; and some of it I was able to answer. That made me feel good."*

It wasn't Feynman. But it was deeply Feynman-like—the inspiring sound of another deeply passionate physicist unpretentiously marching to the sound of his own drum. Or in this case: his own banjo.

The Conversation

I. The Feynman Experience

Inspirational encounters

HB: Tell me more about your interactions with Richard Feynman. I never had the good fortune to meet him, but I'm sure that would have been phenomenal.

DP: It was. He was available, he was cool, he was interesting, he was funny. And it was quite flattering for me, as a young physicist, if he'd come to me with questions.

I often went to him; and there were lessons, there were insights. If you asked him a question, he'd say, "*Wait a minute,*" and then he'd go to this huge wall of shelves with notebooks, the bound ones like you use in elementary school, and he'd pull one out and look at it.

People talk about the great Feynman intuition, but the fact was that he had worked on most of those things in great detail. So when the question came and he didn't remember exactly what he'd done, there was something there that reminded him.

He had a recliner in his office and he'd sometimes close the door and take naps. I once walked passed as he was sitting in the recliner with the door open. And in his lap was what all physicists know of as "The Big Red Book" (*The Feynman Lectures on Physics*).

I didn't say anything to him, but he could tell from my expression what the immediate question in my mind was, which was, *Wait a minute. You're Richard Feynman. You're* **reading** *The Big Red Book? You* **wrote** *The Big Red Book!* This is one of the most famous text books on introductory physics. But it is introductory physics.

He knew immediately what my question was. He looked up and loudly declared, "*It's all in here!*"

HB: What about music? As everyone knows, he used to play the bongo drums.

DP: Well, you would go to the bathroom and Feynman would be there drumming on the top of the urinal as he peed. He was always kind of jiving and tapping.

He'd often say that when he's in a nightclub playing bongos, no one makes a big deal about him being a Nobel Laureate, but the physicists always make a big deal that he's playing the bongos.

The first time I ever saw him was in a film—those famous, filmed lectures at Cornell, *The Messenger Lectures*. There was some trouble with copyright for some time, but Bill Gates recently bought them, because he liked them, and has now made them available online for free.

The transcript still exists as a book called *The Character of Physical Law*. It's a great introduction, meant for interested undergraduates, not science majors in particular. I highly recommend it. A few years ago, in a discussion with Kip Thorne, one of my distinguished colleagues here at Caltech, he told me that he recommends that book to everyone too.

Anyway, in the video of these lectures, Feynman is on stage and there are blackboards on wheels behind him. But the cameraman can't keep him in frame because he keeps going back and forth across the stage so rapidly. He was like an NBA basketball player or something, constantly changing his direction. That's the way he was in real life too.

I remember one day seeing him come down the hall at me really fast, yelling, "*Hey, Politzer, want to see me prove the spin-statistics theorem?*" which is a very deep, important, abstract theorem. And he starts taking off his belt. I was terrified, because I had no idea how far this was going to go and what was involved. But he had this great thing that he did with his belt to demonstrate the theorem.

About a week later he came back to me and he said, "*Politzer!*"— he referred to everybody by their last name, and we referred to him as Feynman; the only one who called him Dick or Richard was Murray

Gell-Mann, and the most deprecatory address Feynman could use was "Professor"—"*Remember that thing with the spin theorem? What did I do? I don't quite remember.*"

I said, "*Well, I think it's like this,*" and I did it. Then he came back the next day and he said, "*You know, I think yours was better.*"

This belt trick captured a piece of the theorem, which was: what's the connection between turning all the way around, in place, 360 degrees, and having two things and switching them? You need some topology or something. There's some profound relationship. You have to think of fields. Faraday invented fields.

I read somewhere that Einstein said the second most important advance in theoretical physics—aside, of course, from the work of Newton—was Faraday's concept of the field concept. Faraday is viewed by many as the greatest experimentalist of all time, but he had only four years of elementary school education, and wouldn't even say that he was self-taught—he never had mathematical sophistication, and fields were something that allowed him to visualize what was going on in a way that had never been done.

One of the giants in physics history is James Clerk Maxwell, who came up with Maxwell's equations. Faraday, who was just a couple of decades older than Maxwell, knew that light was wiggles in the electric field, and Maxwell knew that Faraday knew that. They got together late in Faraday's life and were very appreciative of each other.

HB: I want to get back to Feynman's belt, but first an aside: I recently read the autobiography of Leopold Infeld (*Quest: An Autobiography*), whom I had only heard of before from Einstein-Infeld-Hoffmann theory.

It turns out that back in the 1930s, he wrote a popular book with Einstein. The story is that he was at the Institute for Advanced Study collaborating with Einstein, who tried to get his fellowship renewed, but it was declined by other Institute staff. This was in the late 1930s, which, of course, was not a great time to go back to Poland.

So Infeld went to a publisher in New York and suggested that he write a popular book with Einstein to make enough money to stay in Princeton. Which they did; and Infeld earned enough from the book to stick around the Institute independently and continue his work with Einstein. But the point is that it wasn't just a standard sort of popularization: it was all about the primacy of the field concept in physics, exactly as you're saying, illustrating Einstein's view of its importance.

DP: Good. So I didn't make that story up.

HB: *Make it up?* You're ruining our credibility here, David. Anyway, I do very much want to talk about banjos, but before we get there, what, exactly, was this great belt trick that Feynman did?

DP: I guess I have to show you. Take a belt. Make sure it's not twisted, so it creates a perfect loop when the two ends are connected. Now, hold it in front of you with one end in one hand and the other end in the other hand, and then exchange them. That will create a twist in the belt when you connect the two ends. That twist in the belt is the same twist you would get if you were to take one of the ends, rotate it 360 degrees, and then connect the two ends.

That, at least, is what I told him he told me. I'm actually not sure what he told me the first time. But he said that what I showed him was better. So I'm not even sure it if was different, but that's how I remember it.

The famous theorem, which is one of these important things in abstract particle physics, as we're told by the people who understand the proof—very few people understand the proof because it's **very deep** and **very complicated**—is that it rests on quantum mechanics and relativity. And the question with the belt is, "*Which is which here?*"

Using a belt invokes the field concept, because who cares about turning one end of the belt as opposed to the other? The answer is that the one end isn't just *there*, but its existence stretches out through all of space, and, therefore, these lines, as it were, get tangled in the twisting, which is the same tangling you get by exchanging the two sides. So that's an important piece of what exchanging has to do with turning.

As you know, fundamental particles have two different ways to be identical: they can be identical as you might think, or they can be identical such that when you switch two of them there's a minus sign in the quantum mechanical wave function, and there's a theorem which connects whether there's the minus sign or not, which has to do with how they behave under rotations, which we call spin or angular momentum.

I'm going to tell you one more Feynman story before we move on.

HB: Sure. We can talk about Feynman all day.

DP: Well, these are the Feynman stories that aren't in any of the books, because the ones I tell are the ones when I was *right there.*

Anyway, Feynman was one of the last people to believe that quantum chromodynamics (QCD) was, indeed, a likely description of the strong interactions. He was a sceptic for a long time and had experiments to prove it—he addressed it for a couple of years, but finally he had to admit that that was the way it works.

But the question still remained, *Does that tell us how quarks actually hold together?* Quarks are the substructure of protons and neutrons. So how do you make a proton out of quarks? That remained something that we don't know how to do, and we don't even know that it happens, except that we look in the world and we see evidence that the quarks are there, and evidence that the protons are there, and we have evidence that the quarks are inside the protons, but we can't yet use the equations to explicitly build protons out of quarks.

So Feynman was working on that, because, in his mind, that was the most important thing to work on. He had some ideas. He came in—there was a lot of body language when Feynman was talking physics, especially when the physics was not in a final form—and he gave me the sales pitch of his latest idea.

And I asked him, "*Feynman, do you **really** believe that?*" And he moved towards the door and replied, "*I will quote Emerson: 'In skating over thin ice, our safety is in our speed',*" and then bolted out the door.

It was a real pleasure to have known him, because one of his talents—he had many talents—was to connect things when other people didn't see the connection.

One great real-world example is that he lived at the foot of the mountains here where it gets steep. It was the season just after there had been big fires, where the brush and a lot of the trees had been burnt off, and he was the only one in his neighbourhood who bought flood insurance.

And it began to rain, and the rain ran off instead of soaking in. The streets got clogged with debris and the spillways ran over, and there was water everywhere throughout his neighbourhood. He really was a genius. That shows you how he was able to make these connections that no one else could. Of course that's just one small example. Over a period of decades there were many remarkable things in physics that he was responsible for.

HB: But you never talked to him about music?

DP: I didn't talk to him about music. It was sort of a private thing. In fact, my current research on the physics of banjos began with a course that I started teaching here at Caltech four years ago.

Every decade or so, we review the required curriculum at Caltech. Even if you're not majoring in engineering, it's very much like an engineering school: there are a lot of requirements and the faculty reviews how those work. I was not on the committee, so I don't really know the details, but the way I imagine it went is that somebody said, "You mean to say that students can graduate without ever having had any *fun*? We should require 18 units of *fun!*"

And what I imagine they meant by that is that other schools have this idea of a small class of freshmen matched with senior faculty doing something out of the ordinary that's kind of fun. And the proposal was that there should be a required part of our program that's fun.

So they asked for volunteers for a pilot program. I was one of the first to volunteer because I had always used musical examples and demonstrations in elementary physics: mechanics, waves, electro-magnetism, electric guitar pickups, speakers, microphones—whatever I could fit into the course.

But you can't give a course if it doesn't really fit in, and I didn't think a physics of music course would exactly fit the required need. So I thought that if they wanted a fun course I could do it at a more elementary level, because almost everything that I'm doing now follows on a course we teach to sophomores in the fall term called *Vibrations and Waves*.

But for freshmen the only thing you can count on is high-school physics, and since it's Caltech, they may have a smattering of calculus and their high-school physics is generally pretty good. So if I wanted to create something for them, I had to start there.

Questions for Discussion:

*1. Why do you think that Feynman's book **The Character of Physical Law** is so influential among physicists and non-physicists alike? In this chapter David mentions Kip Thorne as another celebrated theorists who recommends the book widely—in another Ideas Roadshow conversation, **The Power of Principles: Physics Revealed**, particle physicist Nima Arkani-Hamed also spontaneously references **The Character of Physical Law** as one of the most influential books he has ever read.*

2. To what extent are visual analogies, such as Feynman's twisting of the belt to highlight fundamental conceptual aspects of the spin-statistics theorem, necessary to say that we "really understand" a concept in physics? In what way is "mathematical understanding" different from "physical understanding"?

II. Love at First Sound

The joy of the banjo

HB: As you said, you've used musical and acoustical examples for years as a way of demonstrating various phenomena, but presumably behind all of that lies a deep personal love of music.

DP: For me, music is always there, which is to say that there are banjos and guitars on stands at home and here in my office.

HB: How many instruments did you play when you were younger?

DP: The banjo was my fifth instrument. When I was a little kid, I took lessons on the recorder and the accordion for several years. To play the accordion means you need some music theory, because you have the treble and the bass clef, transposition, and accompanying chords, so you have to have a sense of all that.

I was in a boy's choir, which I really liked, but then my voice changed, which was devastating because I couldn't hit all the notes I was formerly able to.

Then I picked up the harmonica. That one I'm pretty good at, which is to say that I've jammed with people and I can actually sound like a harmonica player.

Then came the guitar, which I started playing during what the old guys refer to as "the folk scare". In my circle in high school, everybody played some guitar. And the banjo caught my ear around that time. I built my first banjo when I was 15.

HB: 15? Was that the only instrument you built, or were you also building other types of instruments?

DP: That was the first one. I liked shop in high school. Initially we thought the shop teacher was some mean guy, but in fact he was quite generous: I was allowed to use the bandsaw, which was what I needed to rough out the neck and build the form to make the round drum part. Then it was just a matter of buying the hardware. I was still paying off my guitar and bicycle when I was building it.

HB: How did the banjo sound?

DP: Like magic. That's one of the things about banjos: they all sound like banjos. There is no ideal banjo. Individual people may have their idea about what a banjo should be like, usually based on someone whose performance they fell in love with and they figure that's the right kind of banjo. But there are many different kinds of banjos— and there have been many different kinds over the whole course of history.

They come from all around the world. There are banjos that are indigenous to East Asia and Japan. In Japan it's called a *shamisen*, but it's really just a banjo. There's a classical Indian instrument called a *sarod*, which apparently came to India from Afghanistan. There's the sitar and the sarod. In fact, in my time, the great sarod player was Ali Akbar Khan who was Ravi Shankar's music teacher and father-in-law. The two of them did duets. But anyway, it's a banjo.

So what's a banjo? A banjo is a drum with strings. They came over to America with the slaves. This is a very sad, grim story: it's really worth telling because of how awful it was. Not only was there this trade in human beings, being bought and sold, but the people buying and selling them were worried about how many of them died along the way, because they were stored like cargo. Someone discovered that if you brought along a musician and got the slaves up onto the deck to dance around a bit and then put them back in the hold, the survival rate was greater. Apparently, to this day, there's a group in West Africa who have a traditional tale of the devil coming and stealing the musicians.

Those would have been gourd instruments. In Africa, to this day, there are, as there were then, gourd instruments. There's a skin head, a gourd drum, a stick, and the strings. As I was saying, around the world, there are various versions of that. There are even pictures from ancient Egypt, for example, of a drum, a stick, and some strings.

In America, there are people who play all those different variations and make new ones in those styles. There was no evolution of banjo style that has resulted in an "ideal" or "classic" banjo that everyone now plays. There are actually large communities who play one kind or another. Even more striking is that, if you go to a solo banjo performance, the banjo player will tell stories, sing songs, play the banjo, and will typically come with at least three, but more commonly six, very different banjos. The banjo player might use one to talk about American history, or different eras, or different kinds of music. All the banjos have different voices.

HB: But there is something universal about the banjo. If you were another sort of "string theorist", as some of your colleagues are around here, you might even posit that there is some universal, Platonic form of "banjo" that predated the Big Bang, or something like that: the Big Banjo, perhaps.

DP: Well, maybe. The banjo was chosen because it's loud. I have a banjo here. I'm not going to play it, but I'm going to do one of my favourite experiments. You'll notice that when I rub my finger across the head of the banjo it's much louder than when I rub my finger across the table or the wall.

Banjos are stupendously efficient "transducers", which means something that turns one thing into another—in this case turning small vibrations into sound. If you were to replace the drumhead with a very thin piece of wood, it would look pretty much the same, but it wouldn't sound at all like a banjo; it would sound a lot more like a dulcimer.

I know there are people for whom there is no distinction between a banjo and a ukulele—to them, they're both just popular stringed instruments. But if you listen, you'll hear the difference.

HB: How general is the class of skins that could go over the drum and make that specific sound? Does it have to be a leathery type of substance? I know nothing about this.

DP: Somewhere here in my office is what's called an Appalachian-style banjo, which has a very tiny drumhead. Some years ago there was a great high-school project called *Foxfire*, which was a magazine that ran for several years, in which students went around interviewing local people about aspects of Appalachian culture.

One year they spoke to banjo makers. It turns out that the skin for one of the smaller banjos, about the size of a CD, could be from a cat, or a possum, or what have you. The bigger banjo skin has to be from something bigger, like a goat, or a cow.

Of course, differences in how you treat the skin, thicknesses, and more factors will all result in different sounds. It depends how thin you want it, how impervious to humidity you want it, how white you want it, and so forth. The skins react to moisture in the air and stretch, so the setup of the banjo along with the pressure of the strings will cause the head to sink down, and sometimes you can't play it, so you have to put a candle or a light bulb inside to warm it up before a performance. If you tighten it up and the weather gets dry and you forget to loosen it, it will crack.

In the 1950s the idea of a plastic head was introduced. If you ask the professional performers of that era, they'll tell you that they hated the sound compared to the skin heads, but it was a convenience, so they made the conversion. The people who fell in love with their music now wouldn't play anything other than the Mylar ones because that's what they were used to hearing.

I learned how to mount a skin head. It's not actually leather; it's like a hard cracker. You soak it and it gets soft and becomes stretchy. You bend it, let it dry and it shrinks and tightens up. I read a long article by Sam Stewart, one of the great popularizers of the banjo who had the biggest banjo company in the world in Philadelphia around 1890. Anyway, this article describes how to mount your own skin head, and in the end he says that it's kind of a crap shoot, because each skin is a little different.

There's a ring around the head that tensions the skin and pulls it down, like a drum. You can just stretch it and nail it in around the edge, but then you really have to know what you're doing, because there's a limited amount of play that you have in terms of tightening the ring down and stretching the head because of the neck. So you might find it's either too tight or too loose because of how it shrunk while it was drying. Experience helps. If you're doing it again and again and again, you'll likely be much better than someone like me who does one every couple of years.

HB: When you were 15 and you made your first banjo, were you thinking about how you could get one particular sound as opposed to another, or did you just want to make a banjo and see how it sounded? That is, were you thinking at all of the physics of the sound at that point, or did you just think, *I want to make a banjo and see how it sounds?*

DP: I think it was more the latter. But I'm kicking myself now, because there's a phenomenon of plucked strings, which is visible to anyone who is vaguely observant. I observed it then, but until a year ago I had never asked, *Wait a minute, what's going on there?*

When you pluck a string, the vibrations of the string would generally get smaller and smaller until they die out. But sometimes the vibration sort of pulses, and you can see that. What was going on? I didn't pursue it. If I had, and understood it, I'd have made a big breakthrough and been famous. Anyway, we'll get to that later: coupled, damped oscillators. That's a big story.

But when I was 15, I just played it, and it made banjo noise right away.

HB: What was it that particularly appealed to you about the banjo sound? I'd like to talk about characterizing the sound itself before we get into the physics of that sound and what that actually means.

DP: That's an absurd question to ask a banjo player.

HB: I specialize in absurd questions.

DP: There's a definite charm about the banjo, there's something very special about it. Pete Seeger, late in his life, stuck it into every interview he did. He talked about the wonder of those, what he called, "little points of sound", like stars, lots of them, and out of the zillions of them—any kind of banjo playing tends to have a lot of them—the melody emerges.

For many people my age, their first banjo instruction—if you didn't have somebody you knew who was teaching you—was a book written by Pete Seeger in 1948. It's still in print. It's called *How to Play the Five-String Banjo*. I remember an interview he gave when he was in his late 90s saying that if that had been the only thing he had done in his life, he still would have had something to really be proud of. He did a lot of great things. He was a hero to many people.

HB: I would really like to hear you play. I imagine that there will be people watching this who think this is all very abstract; and I know that you, as somebody who has made profound contributions to theoretical physics, nonetheless believe very strongly in the empirical world and being concrete. Surely you recognize the irony, if not the downright inappropriateness, of talking about banjos without actually giving people a taste of the sound of one. Would it be possible for us to hear you play?

DP: Well, I'll think about it while I'm answering—I have to say that my banjo work is different from much of my other scientific work. I've published in scholarly journals for my physics career, but when it comes to banjos, the few things that I've read about other people's work I didn't find in journals, but just by googling "banjo physics".

The things that I first posted had sound samples, because that's essential. It had measurements, but you don't really want a sound sample of the pluck of a single string. In the end, the question is, *Can you hear the difference between this banjo and that banjo?*

I've built a few over the years. At some point, I got hooked into the idea of buying old ones and restoring and repairing them and that got really interesting. There are a lot of design decisions that have to be made because there is no real standard that you're going after.

HB: You're stalling, aren't you? You just don't want to play the banjo in front of the cameras.

DP: I could.

HB: See, here's what I'm thinking. You're telling me about these points of light. You're telling me that I'm asking absurd questions about why anybody would want to play the banjo. Fine. I'm not a "banjo guy." I've never really thought much about the banjo before. But I want to have a banjo epiphany.

DP: All right. Fine. Well, I'm not stalling, but I should tell you that people ask me to play and inevitably they're disappointed. It doesn't sound like Bela Fleck or Earl Scruggs.

HB: Well, OK. I understand that you're not the greatest banjo player in the world. That's not why I'm here.

DP: I do enjoy playing it, and when I get together with serious musicians, I have no problem playing.

That reminds me of a good joke. Technical people are often concerned about how you mic a banjo. It turns out that it's one of the worst instruments to mic, for physics reasons.

HB: Why is that?

DP: The directionality is highly directional and highly frequency dependent until you get to the far field, but then you get room sound, and floor bounce, and so forth. So you come close, you go far, you mix them—if you go online and ask what the best distance is, you'll find that about two miles is a good distance.

HB: That could get a bit complicated with our present set-up.

David then plays the banjo for a short time.

Questions for Discussion:

1. In what ways does making an instrument give one a deeper appreciation of it?

2. Why do you think that most of the work on the physics of banjos lies outside of mainstream scientific journals?

III. The Holy Grail

The challenge of quantifying sound

HB: I'm starting to get a sense of these pings that you were talking about, as well as the use of the drumhead which you're regularly hitting while playing. Presumably some guitarists do that as well, but it really comes out more effectively with a banjo.

DP: Well, there are a lot of different styles. There are guys who plant their hand on the drumhead and don't hit it at all; they just pick.

Another feature that some banjos have is a so-called "drone string". It's not a worldwide feature, but it does exist—and it existed in the African instruments that came to the Americas.

HB: What's a drone string?

DP: It's this shorter string on the top of the banjo. Like the bagpipes, the drone string is pretty much constantly playing the same note while others are being played simultaneously. It's rarely fretted, which means that the frequency never changes. There are some people who do fret it. Anything is possible. There's nothing they don't do. Long before Jimi Hendrix, performers were playing banjos behind their heads and throwing them in the air—although to the best of my knowledge they didn't set them on fire.

The banjo is also reentrant, like the ukulele, which is to say that, unlike the guitar, say, the pitch of the strings doesn't continually increase.

There are also many different tunings. Almost everyone who plays a violin in a concert hall tunes it the same way. Fiddlers for dancers have their own tunings—they're like buddies of banjo play-ers—and banjo players have a zillion different tunings.

Because going from one tuning to another can take a while, the famous banjo player Earl Scruggs invented a gizmo which allowed him to change tunings instantly. Then he got really clever and started to do it in the middle of songs.

HB: Why would anyone do that? First of all, why wouldn't someone keep the same tuning throughout?

DP: Lots of reasons. A simple reason is that the open strings tend to ring better than the ones you fret. So, depending on the song you're playing, you'll want certain notes to ring.

There are other reasons too. Right now, for example, I'm in a tuning I particularly like, in which the drone is not the tonic but the fifth. The tonic is the first note of the scale. The drone in this particu-lar tuning is the fifth note. There are other tunings where the drone note is the tonic, which has its own unique sound.

In my class, the students are interested in a scientific explanation of the things we know and like about sound. Octaves—the same note, just eight notes higher—are something universal to human beings. You can even train animals to recognize the sound of an octave.

If you play a note—I've done this with my students—and you ask the women to sing it separately from the men, you'll typically find that they differ by one octave. They heard the same note and they think they're singing along, but they're singing an octave apart.

Anyway, when you divide the string in half, you get an octave; it's double the frequency. When you divide it into thirds, you get a fifth.

When you pluck a string, it's doing a lot of things: in the first case it's just going up and down, but it also kind of wobbles a little, which creates the harmonics. The octave and the fifth are common to all human music cultures—I'm told you can train animals to recognize fifths too—but after that, you're on your own. The way you divide the notes within the octave, besides being sure to include the fifth, depends on the culture.

One of my banjo heroes, Jens Kruger, came to the United States from Switzerland. He's a dynamite bluegrass player, but he plays all sorts of stuff. There's a recording you can find on YouTube of him playing Bach's Cello Suite No. 1. It's awesome. When you listen to it you think, *Bach must have written music for the banjo*, because it's eighth-note arpeggios, which means playing the different notes of a chord separately, rather than simultaneously. And it goes up the chord and down the chord, and then another chord.

HB: Moving towards the physics of the banjo sound, let me ask a basic question. What is it that makes one banjo sound different from another banjo?

DP: I wish I knew. That's really the Holy Grail. That has fascinated me over the years, especially when I got into fixing up old banjos, seeing how different they were, and picking out what I liked about each different sound.

HB: As you know, I had an Ideas Roadshow conversation with Joseph Curtin about his work of objectively comparing the sound of old and new violins (*The Science of Siren Songs: Stradivari Unveiled*). There's this mythology about Stradivari and Guarneri violins, a general belief that this represented the apex of violin making that could never be repeated. But it turned out that it was actually non-trivial to understand what one meant exactly by the sound, to characterize the sound. And when violinists, violin makers, and other specialists were involved in various double-blind experiments, it turns out that they couldn't actually tell the difference between the old, iconic violins and the new, modern violins. Is there anything similar in the banjo world?

DP: Well, before I answer that, let me add one point: they couldn't identify the instrument, but they **could** easily tell who was playing. A *really* profound challenge that I face when I record banjos and try to understand them is, *What is it in the sound that we identify?* Now, with the Stradivarius violins, one of the problems we have is that we identify the price tag. Not only could you not tell the difference by looking at the voltage trace as a function of time or looking at the spectrum analysis using your computer, but the people can't tell.

I had the pleasure of talking to the Music Director of the Pasadena Symphony recently. We talked about instruments. I was interested in the mechanics, and he knew a lot. He talked about using the different violins and accommodating the differences—because there *are* differences. We *hear* differences, but for me it's really fascinating—and really frustrating, because of how hard it is—to understand those differences in a measurement sense.

It's like voice recognition. It's not just about identifying the word, but it's also a question of recognizing *who* is speaking. **We** can do that: you pick up the phone and you know from the first word who it is—even though the sound may be somewhat distorted through the phone, you still know. You may have trouble recognizing recordings of yourself, but that's another story. We recognize other people, snippets of songs from just the first couple of chords, and so forth. *What is it that we're doing there?* This is not easy.

There are banjos with very different sounds, and they're made differently. The parts are very swappable. The banjo I'm holding now came with a different head, bridge, and tailpiece. They each make a difference in the sound, but I don't know where to point on the recorded trace of the voltage as a function of time of the microphone to show where those differences are.

Those are the kinds of questions that I'm trying to solve. If you've got a physics story that's trying to explain the different sounds, it's going to have equations and predictions for numbers; it's not going to tell you that the sound is "warmer" or "tubby". That doesn't come out of an equation. It's a question of the relative strengths of some frequencies compared to others—and there were a couple of cases where I thought I was able to make a little forward progress.

Questions for Discussion:

1. What do you think it is about octaves and fifths that accounts for their universal nature amongst both humans and animals?

2. To what extent can we be certain that different people are "identifying" the same fundamental characteristics of the same sound?

IV. The Ocarina Effect

Probing the effect of rim height of the open-back banjo

HB: Let's talk now about some of the specific issues that you've worked on. You mentioned coupled oscillators earlier. I'd like to get to some things that have captivated you, not just on a musicianship level, but also in terms of your interest in physics and figuring out how these things work.

DP: Well, the first one has a good story because it has a physics end, with lots of physics along the way, but it's also what really got me into the field.

I mentioned earlier that I teach this freshman seminar on music. I have the students do a project of their own choice and design, inspired by the stuff that we've read and talked about in the first half of the term. They've done very interesting experiments on topics like human subjects testing sound perception.

The first reading we have, Dan Levitin's *This is Your Brain on Music*, talks a lot about that. It's a good introduction to that kind of thing. Some students have made instruments. That's what I like most. They build a simple instrument and then listen to it and say something about how it works.

So I thought that I should do a project along with them and share it with them. I wondered, *What should I do?* Well, I'm a big banjo fan, so a natural subject would be to examine the sound effects that design decisions on a banjo are going to have. Many of those I already know. For example, one you might not think of that's kind of subtle is that the choice of wood for the neck is really important.

HB: Why?

DP: Well, let's examine the way the string vibrates. One end is at the bridge, and the other is at the nut, just below the headstock. The bridge is vibrating and talking to the drumhead, and that makes sound. At the other end the string is vibrating and pushing up and down on the nut. The question is, *Is the neck perfectly rigid or does it have some flex?* There's quite a range of woods. The typical ones are really stiff in terms of the vibrations, and they range from maple to something like walnut or mahogany, which is softer. In that case, the highest frequency pieces of the vibration just get lost into the neck. They don't come back to the drumhead. So the choice of wood for the neck has a real impact on the sound.

There's a long list of those sorts of design details where I know what effect that decision is going to produce: different drumhead materials, thicknesses, tensions, and so forth.

There are two big classes of banjos in America today. There are the ones that bluegrass players play, which have a back that wraps around to the front of the drumhead. Now, on an acoustic guitar there's usually a round hole on the front of the guitar.

Violins have the F-holes, also on the front, whereas with the resonator banjo the hole is created by an air gap between where the back wraps around the banjo and the rim. The size of that hole relative to the size of the air inside the drumhead determines the lowest body resonance of the instrument.

Strings by themselves don't move much air and therefore don't make much sound. The signal from the strings needs to be close to a resonance, in this case the drumhead. It's like singing in the shower when you hit a particular note and it gets really loud. The lowest sound vibration comes from the cavity and the opening, and is affected by their relative sizes. Anyway, that's the resonator banjo.

The other type of banjo is the open-back banjo, where there's nothing there. And the question I had was, *How do you decide how high to make the rim of the open-back banjo?*

I knew a physics answer for why you don't make it too thin: because it will sound thin. In audio systems, you put woofers in a box because the woofer is pulsing to make low-frequency sound, so you're squeezing the air in front while expanding the air behind. For the very low frequencies, the behind sound comes around the side and combines with the front sound and cancels.

Basically, you have to eat, or use in a clever way, the sound coming off the back, because otherwise it's going to come around the front and be 180 degrees out of phase by the time it reaches your ear; it will be compressing while the other is rarifying. They arrive at your ear at the same time and cancel. So it sounds thin. You lose the low notes if the banjo is thin or doesn't have a box.

I also know that the rim of the banjo can't be higher than about a foot, because it would be very hard to actually play the instrument. But why is the rim exactly the height that it is for an open-back banjo?

I wrote to the four people in the universe who had published scholarly papers about banjos, only two of them in a proper journal. I also wrote to another physicist named Eric Heller, who is at Harvard. He's a physical chemist and physicist with great insights about quantum mechanics based on understanding waves in a visceral way. For several years now he's taught a course at Harvard for general undergraduates and he wrote a great book called *Why You Hear What You Hear*. It's very ambitious for an introductory book, and it covers lots of interesting things.

I wrote to all these people, and they had no idea. In retrospect, I know why: because they imagined this poor banjo in a laboratory: some researcher and his undergraduate co-worker examining a resonator banjo clamped down on a lab table while they plucked it with a carefully calibrated plucker and measured the sound with sophisticated instruments.

But that's not how it's played.

It's played, as you just saw—for better or worse—against your belly, or maybe resting on your lap. And that makes a difference, because having your body behind the banjo creates a lower tone. Like all stringed instruments, there's an enclosed volume with a sound hole. I can make the sound hole bigger or smaller by moving the banjo farther away or closer to my body.

There's nothing new under the sun. Good players know this, and they use this technique. I remember watching a video of a young guy who was talking about giving lessons, advice, and workshops for people, and he said, "*Before you go out and get another banjo, consider playing the one you have in different ways.*" He was talking about open-back banjos, holding them differently, and the very different sounds that you can get as a result.

The volume enclosed and the size of the hole determine this lowest resonance, called the Helmholtz resonance—it's the sound you get when you blow across the top of a bottle. You get a much lower note than if you had a straight pipe with a closed end. If you closed the top end and blew across the open one, you'd get a higher note. There are a lot of interesting ideas related to this.

Anyway, I had banjos that were similar with slightly different heights, but I realized, through recording them, that the discernible differences in sound might well be attributable to their other minor differences, like where the bridge sat on the drumhead, for instance. Once you have frets, you can't move the bridge around.

So my wife said, "*What you need is for someone to build you identical banjos, except for the rim height.*" Now, there are many people these days who build individual banjos. They build wonderful instruments. But this seemed far too extravagant for me to undertake.

But it dawned on me that there's one guy in the world who might be able to help. He's in San Diego, and his name is Greg Deering. He was an industrial arts wonk who went to college to study industrial arts to build banjos, but the professors scoffed at him, so he dropped out and just started building his own banjos. His goal was always to build a quality instrument at an affordable price, and along the way he learned that those are both relative terms.

About 15 years ago, he had a vision of doing much better at the affordable-price end than he'd ever done. Recently he passed the 100,000 banjo mark and he now has 45 employees. For a long time it was just him and his wife and a small shop. Now he's at the point where he can't meet demand.

The banjos he makes retail for $400. I had a small research budget. I went to my department chair and said, *"Could I use that for banjos? It's related to my class."* And he agreed. Three banjos was the minimum number to properly test the different ideas.

So I wrote a letter to Greg Deering. I thought maybe I could get them wholesale. And he replied, *"Come down to my place and we can talk about it."* I told him that I was a college professor, a physicist, and that I was interested in banjos. He thought that was pretty cool. So he gave me a tour of the whole factory. He was really proud of some of the machines that he'd invented, often recycled from old machinery, and the industrial production processes that allows him to ultimately sell this banjo for only $400. His estimate was that it takes three hours and fifteen minutes to build a banjo, from lumber to FedEx truck.

HB: Really? I'm astounded.

DP: That's what you're supposed to say. It takes me that long to figure out which saw I should use and maybe go to the hardware store to get a new blade.

HB: My immediate response is that doing it that way must cut down on quality somehow.

DP: No, they're marvellous instruments.

Who buys them? Music teachers will recommend them because, out of the box, they're perfect. There are professionals who perform with them. Others will say they love them because they are light and they can travel with them. It's a professional-grade instrument. It has a different voice than heavier ones.

HB: When you approached him and got the tour of the factory, I'm guessing that he was not only excited by the fact that you were a college professor who was interested in banjos, but also the specifics of your request, right?

DP: Yes. After he showed me his factory he asked, "*What can I do for you?*" So we sat down with pencil and paper. It was really cool because I had some ideas of what the minimum and maximum heights that I needed were, and I had originally imagined buying three banjos and modifying them myself. He pointed out that there would be some particular issues that would need to be accounted for. For example, if one were to be lower in height, the mounting of the neck and the tailpiece would have to be different. He had ideas for how to do that. So all three were built that way.

He ended up saying, "*It's on me.*" The three banjos were as identical as can be, except for the differences in rim height. Someone who makes banjos one at a time could not make them that identical, because these ones came off of the computer-driven mills, were checked for quality control, and so forth.

He has also made improvements since he began his process. I first thought that the changes were solely aesthetic until I played a new one, but the sound was even better. They went from very good to really terrific.

HB: So you got these three banjos and you were able to do some experimentation.

DP: Yes. And along the way it became very clear why the pitch changes as you open up the banjo away from you as you play it. Do you know what an ocarina is?

HB: No.

DP: Ocarina is Italian for "little goose". I have one in my office here. It's a spherical object. It's like a bottle you blow across. The concept is that you have the same volume, and as you increase the hole size, you get a higher note. That's how an ocarina works.

And that's basically what I demonstrated with the open-back banjo—the change in tone when I moved it farther away from my body corresponds to increasing the sound hole which results in an increase in the pitch. Correspondingly, if you make the volume bigger for the same hole size, the pitch goes down. There's a formula for that which goes back to Hermann von Helmholtz, who's one of my heroes.

In short, the solution to the mystery is that the open-back banjo does not really have an open back, because you have to explicitly take the way the player holds the banjo into consideration and how closely he holds it to his body.

Recognizing that was relatively simple, and fitting the measurements to equations—well, I was kind of proud of that. It kind of worked. They sounded different when you played them. There's also the issue of where you sit when you play, because the player doesn't hear the banjo the way the listener does. They're very directional and very frequency-dependent on the directions as I said earlier, so it's best to have somebody play it for you. I went to a buddy who plays much better than I do and we tried the different banjos. He liked the big fat one.

Questions for Discussion:

1. Would it be possible to hold an open-back banjo in a way to make it sound roughly equivalent to a resonator banjo?

2. What do Greg Deering's production methods imply about our standard beliefs of the superiority of "handcrafted" techniques over automated?

V. Hearing Pitch

Not so simple

DP: The thing that's simple in terms of physics is not the whole sound of the banjo—and in fact I don't really have a story, a physics story, to explain the part that my friend liked—but in terms of physics, what you can say is that a banjo has a richer bass sound because it resonated at a lower note, it's lowest resonance: the lowest note that it could actually make when you measure it.

What's very interesting about the banjo is that, as you would expect, as you play various notes that descend in pitch and analyze what sorts of frequencies are in that note, the strongest frequency will be the one that you identify with the pitch.

But then, as you descend, at some point things change and what you hear is actually a combination of other frequencies, but the frequency that you identify with that note isn't actually there. It's kind of weird.

There's a little bit of arithmetic involved: it has to do with Fourier analysis, and the sine waves, and all that stuff. If you have a sound that is a combination of something that is 200 hertz and 300 hertz and you start them at the same time; that will repeat every 100th of a second. You will hear 100 hertz, but, in terms of the frequency spectrum, there will be no Fourier component—there's no sinusoidal piece of the wave at 100 cycles per second.

What's happening is that the drumhead doesn't vibrate at the same frequency as the note that you hear—the string is vibrating, but we're not making sound with that sinusoidal wave. But we've got enough of the other ones, the harmonics, for us to somehow interpret a pitch.

People debated this, and I don't know if they've settled the issue. I got my point of view from that book by Eric Heller that I mentioned earlier, *Why You Hear What You Hear*. He said, in effect, *"Let's not look at the Fourier spectrum. Let's just acknowledge that it is periodic,"* and there's a name for that: the auto-correlation function.

Our brain is smart enough to do many things at once. Part of the distraction for a physicist or acoustician is that there is a piece of our ear that identifies pitch, called the cochlea. It looks sort of like a snail, and when you unroll it—it's long—it has a tapered membrane inside. And when you send sound in, it vibrates a lot at the fat end, to the extent that you send in low frequencies; and it vibrates a lot at the thin end, to the extent that you send in high frequencies. So it's an analog Fourier analyzer. And you have nerves along the way, stuck on the bottom, and each nerve knows where the membrane is wiggling and how much, and it sends that signal to the brain.

HB: So the idea is, as I understand it, that we're trying to understand why something sounds the way it sounds, and to understand the way something sounds we have to look at what our own detection device is actually doing internally. We have to understand the physics and the mechanics of the inner ear and how our brain is processing the information and so forth.

I would have naturally assumed, out of ignorance I suppose, that you don't need any of that: you just take the sound, and you're able to process it using whatever—

DP: Exactly. Physicists are very attracted to the idea that sound is going to be simple. Okay, we have two ears, but we can pretty much hear with one. So one ear is like a microphone and a microphone responds to sound by giving you a voltage as a function of time: a real variable with some finite range that is a function of time. That's the whole thing, we figure.

Yes...but our brain—as the computer guys will tell you—is massively parallel. That's to say the rate at which our brain processes bits is something like a thousand or a few thousand per second.

Our computers do billions per second, but our brains are a lot smarter because they do a lot of them at once and put a lot of information into something like a "buffer".

That is, it doesn't look just at a function of time; it looks at a whole stretch of time and then asks different questions about it and makes its decision about what it's hearing and how to interpret it based not just on what's happening right now but rather what's happening over a period of time and comparing inputs. It's doing all these different things simultaneously and reporting back.

Neurobiology is terrifically fascinating. In the class that I give, we touch on that, and I go as far as the students are willing to go. The first book we read, *This Is Your Brain on Music: The Science of a Human Obsession*, was written by a guy named Daniel Levitin. He used to be a rock musician, then became a recording engineer, and then went back to school because he became fascinated by this question. He does the neurobiology of sound and sound perception.

HB: OK, so to me, at least, there are two issues at play here. One is how we process the information with all of our neurobiology, and the other is this notion of what we're actually hearing—and they're clearly related. If you're telling me that in my ear I am identifying specific signals using this analog device that receives these things—if I don't appreciate that and I'm just looking at the mathematics without any awareness of my neurophysiology—I might assume that I don't have any selection procedure for one frequency over another because of the physiology of my ear.

Which is all to say that that's also a factor that I hadn't really appreciated. I mean, it seems obvious: we're humans, we're listening to things. But when we ask questions about what sounds good, or what something sounds like, or how something can sound the same as something else—it's clear after a moment of reflection that this must be predicated upon my neurophysiology. That's something that I didn't consider, at least with respect to our understanding of sound.

DP: We do marvellous things. Your recording guys know this. Take room sound. If you record something, generally you can tell where the recording was made; but if you're there while the recording is being made, to a certain extent, you become oblivious, because our brains are good at focusing on a specific piece of the information.

An example of a visual parallel would be when you look at a wall and it appears to be painted off-white. But then, if you look at the wall using a light meter, you'll find that, not only is the intensity radically different as you move along the wall; the colours are different too. The spectrum is different. But we see it as just a white wall. These are optical illusions.

With sound, not all the information is in the microphone, because we do some other amazing physical things. Our ears are complicated shapes and they reflect the sound down and in to our brains in some complicated way. I don't know much about how head motion affects our perception of sound, as the vision researchers have been doing this kind of stuff for images for much longer. As a physicist, you might just think you've got a lens and a screen—the retina—and you project the image onto the screen, and that's what you're seeing. That's **totally** wrong.

Not only do your eyes move from side to side; they rapidly shake back and forth in minute amounts. It's as though your camera man were a speed freak, shaking a handheld camera back and forth. That's the image on the retina, but what you see is this very calm, steady world.

You are creating this whole, sort of, Pixar thing where you're taking the pixels that are projected towards you and you're creating some Disney cartoon, which is a whole other reality, based on the initial, received information. You're not image-stabilizing in the sense of compensating for the motion of the camera: you are creating a concept.

Sound has aspects like that. It has to do with what else is going on, what we're used to, what's familiar, and so forth. The course I teach gets into the subject of art—we want things that are both familiar and challenging, and some balance. That depends on our past experience and how ambitious we are.

HB: Do you think it might it be the case—perhaps not now, because we're in the preliminary stages, but in 50 or 60 years when we can start harnessing some of the empirical evidence that we're starting to gain from neurophysiology—when we might start being able to somehow apply our understanding of sound in a different way, thereby perhaps going the other way, as it were?

DP: Maybe. But I'm a Luddite. You're asking the wrong guy.

HB: Well, what do you think?

DP: I'm a sceptic, and I'm definitely not an enthusiast.

Musical recordings go back about 100 years or so, while the oldest identified, archaeological, musical artifacts are about 50,000 years old. And presumably people were making music long before that.

HB: Really? 50,000 years old?

DP: Yes, somewhere in central Europe they found flutes made out of bones. You can tell because somebody put the pitch holes in the right place. Earlier instruments, like drums and stringed instruments—you're not going to find clearly identifiable objects.

Let's imagine that music is older than that. But it was only 100 years ago that we started recording and transmitting, and that has changed humanity forever. The great thing is that we can hear old recordings—they were on these wobbly 78s, collectors found them, and now you can get a CD of them.

On the other hand, I'm sure that, per capita, we do a lot less music-making than we used to. And I think that's for the worse.

I don't think people are very different from how they were hundreds of years ago. I don't foresee anything that we're going to learn that will allow us to vastly improve on things. Maybe I see too many downsides, I don't know... I guess there might be some upsides to knowledge about how people think and work.

Questions for Discussion:

1. What do you think David means, precisely, when he talks about how our brains are "creating a concept" of the world around us? How does this differ, exactly, from the naive belief that the brain "passively records sensory inputs from the surrounding environment"?

2. How do you think moving our heads would affect our perception of sound?

Those who are interested in exploring more about how our brains rapidly integrate a variety of different sensory information on an ongoing basis are referred to the Ideas Roadshow conversation *Knowing One's Place: Space and the Brain* with Duke University neuroscientist Jennifer Groh.

VI. Relative Strengths

Break angles

HB: I'm happy to have a discussion later on about how we're all going to hell in a handbasket, how technology is pernicious because it stops us from enjoying simple pleasures in life like playing music, and how instead we're all running around looking at our phones and bumping into things—

DP: You see—there you go. You get it.

HB: ...but first I'd like to move on to some of the specific work you've done, shedding some more light on the physics of banjos that are counterintuitive or that we might not have thought of before, finding out how things actually work.

Earlier, you were talking about the effects that the vibration of one string has on nearby strings. You've also looked at how moving the bridge up might affect things.

DP: There were a couple of projects and they sort of worked in different ways. They were very satisfying in the end, but they were very confusing along the way.

HB: OK. So tell me in a little more detail about what was so confusing.

DP: Well, I'm not interested in questions I already know the answer to. On the other hand, if I don't know the answer, finding out from someone else is a pleasure, but figuring it out by yourself is truly a wonderful thing. Often someone else has been trying to tell you the answer all along, but it still feels wonderful if you've had a chance to really think hard about the problem. You may not realize that you finally understand what they've been trying to tell you all along, but that process is still wonderful.

Sometimes no one has actually been telling you directly, but then you discover that a lot of other people already knew it. Then, every once in a while, you find out that this thing that you were glad to discover is something that nobody else knew, or most people didn't know, and then they make a big fuss about it, depending on what it is.

Often it turns out that somebody else knew anyway. To understand is a profound joy. It's also a mysterious thing: *what does it mean to understand?* I've come to realize that different people have different criteria for that.

Anyway, there were a couple of things that I didn't understand in the beginning, but thought I should—which, for me, makes it something worth pursuing—and then, in the end, I thought I understood at least something about it.

I told you before that I made these fat banjos, which my buddy liked. And he had helped me so much that I decided that I would make him one too. So I bought a Deering banjo and I made it fat. But it turned out that he preferred a different tailpiece, one that I had used on some other banjos.

Frankly, I had never paid attention to that specific aspect of a banjo. I knew that there were different kinds of tailpieces, and that sometimes people made adjustments, and that some people made a big deal about them, but *I* never had. So I read what people had to say about the different kinds of tailpieces.

Most of this information came from magazine articles, and none of it made sense to me in terms of physics equations. I couldn't turn the words, which were a story, into something that had to do with the force of the strings on the bridge, the force of the bridge on the drumhead, and how the drumhead moves.

So trying to write equations for it was very difficult. It's called the break angle, the angle at which the string bends when it goes over the bridge and down towards the tailpiece. And the claim is that the break angle is responsible for making a banjo "banjo-like"—more of a metallic sound, more "ping" and "clang"—whereas another design with a different break angle will be "mellower".

Lots of sources say that. If you go to the Deering website, there's a great guy there who's responsible for checking each instrument before it goes out the door. He's in the sales department and he's also a great musician. And on their website, he'll give you some tailpiece advice, and that's exactly what he says.

But for the life of me, I was having some trouble putting that into some equation model. Then I ran into an issue—if there's a bend there and the bridge is going up and down, the string has to stretch.

Now if you go back to physics that you learn as a physics major, you'll recall that the shortest distance between two points is a straight line, so if the string is vibrating, it's not as short as it used to be: it had to stretch a little bit to go sideways. But we have a long song-and-dance about why we can get away with ignoring that, which is what we do when we get equations that give a really good description of the sound of the string.

So all this business about harmonics that incorporate a basic description of how stringed instruments work ignore the fact that when the string vibrates sideways, it stretches. Now, if you go sideways a very small amount, the stretch is even smaller still. Meanwhile if the bridge goes up and down a very small amount, the stretch is of the same small order.

So I had the sense that maybe this effect has to do with making the string stretch while the bridge is going up and down. When you tighten or loosen the string, the pitch goes up and down. So somehow it's like tightening and loosening the string while the string is vibrating.

At that point, I stumbled on something while researching online, which was a description of a discovery made by a professor of music at Stanford. He's a composer who's interested in digital music. He made this discovery, patents it, and it turns into—according to the group that he works with, at least—Stanford University's second biggest moneymaking patent of all time. His name is John Chowning. They went into a collaborative R & D effort with Yamaha instruments and produced the first generation of consumer electronics with a keyboard where, when you press a button, it doesn't sound like an electronic instrument—like a Moog—it sounds like a *real* instrument: a clarinet, or a banjo, or a guitar.

What did he discover? He discovered that, when you stretch a string, the pitch goes up. I never know whether it's called tremolo or vibrato. I've asked people and they don't know either. But anyway, what Chowning found is that he could increase the frequency of the tremolo or the vibrato to a point that it was so fast that you couldn't hear the pitch going up and down. And that causes the tone, the timbre of the note, to become metallic. When I first heard that, I thought that might be the key to the mysteries of the banjo sound.

HB: How do you physically profile what you mean by "metallic"?

DP: Well, that was a heartbreak. I haven't a clue.

All I know is that I can take—just like Chowning did—a computer-generated tone and give it just a little frequency modulation and it will sound almost bell-like. And I also know that the tiny up and down movement of the bridge is causing the same order of string stretch. That's what I had to work with, but professional acousticians were unimpressed, because in their eyes I didn't comprehensively settle the whole issue.

This was pretty odd to me because I come from a different field. I wrote a paper when I was still a grad student where I did a calculation. I knew the calculation was right, because I checked it. And I suggested that it accounted for a particular phenomenon, which was something of current interest, and it took the whole physics community, including the big accelerators and whatever, about four years to decide that it was right.

But it wasn't incumbent on me to do **everything** to show that it was ultimately correct. In acoustics, on the other hand, I was somehow expected to settle *the whole deal* single-handedly.

As I mentioned before, there were maybe two published papers on banjo acoustics. These guys work on violins, guitars, pianos.

There's a literature stretching back eons. So if you have some new idea, it has to fit in with all those techniques and all that knowledge.

In short, then, this one is a little bit of a heartbreak. I think perhaps my biggest contribution is to draw attention to the instrument, which I think is charming, and to say that there is something common to the sound of all banjos, by which I mean drums with strings. It could be a gourd with a goat skin and gut strings. It could be steel strings, Mylar top, and a resonator back. They all sound different, but they're all recognizably banjo. If they're recognizably banjo, then the accounting of that aspect of their sound has to be in the fact that it's a drum with strings, because that's all they have in common.

HB: Might there be some phenomenon that is analogous to this with the piano or the guitar? Might there be some connection between what you've found and instruments that aren't banjos?

DP: I tried to put that in my paper. Other acoustic stringed instruments have some version of this, but I can tell you why it's much less important, and that's related to the fact that the banjo is loud because of the resonating drumhead. For a given force applied to the string of a banjo, you get much more volume than you would get on other acoustic stringed instruments. There's much more up and down motion of the bridge.

Now, some other instruments have a bridge with a break angle. With flat-top guitars, for instance, when the bridge goes up and down, the string goes up and down with it. It's the fact that the tailpiece is fixed on the banjo, while the bridge is going up and down, that makes the string stretch.

HB: You know what this seems analogous to? Maybe I'm overstating things, but I'm reminded of this whole idea of when Kepler was looking at the planets and what shape they were. Everyone thought they were circular—

DP: Wait, no they didn't. Ptolemy knew that there were epicycles—

HB: Sure. Circular combinations—

DP: Well, that's called Fourier analysis.

HB: Hold on. I'm going for an analogy here. You can tell me it's wrong or inappropriate or whatever when I'm done, but first let me finish.

So the analogy is that Kepler is looking at things and it so happens that Tycho Brahe tells him to look at Mars. He's looking at Mars and, lo and behold, although it takes him seven years or whatever, he eventually realizes that it's an ellipse. But he's able to make that discovery because Mars has the greatest eccentricity of the planets that he might have been looking at. Had he looked at some other planets, he might not have actually noticed this. So that's my point.

So my somewhat tortured analogy is that you're saying banjos have these effects because of their particular banjo-like nature, but perhaps analogous effects are smaller and harder to detect, or somehow not as meaningful, in other instruments, but it's still enough to draw *some* conclusions and recognize the importance of a particular phenomenon or relationship. You see where my analogy is going?

DP: Yes, but I don't like it. There's this long tradition of physicists—it's not all physicists, by any means—who get hooked on music because it's important to them personally, and they see it as culturally important. If that's what you're doing, you have to make contact with the music as music. There should be some aspect of that.

I'll give you an example from my readings for the course I give. My favourite example is a comparison between the clarinet and the oboe. They sound different, and the question is, *Can you account for the difference in sound?*

I won't even try to explain the physics right now—it requires a lot of diagrams—but the important thing is that it's not due to the effects of the single versus double reed—that turns out to be a *consequence*, not the *cause*, of the difference.

The deal is that most of the clarinet has a straight pipe, and almost the entire oboe is composed of what's called a conical bore: it's a fixed-angle thing. That's why it has a double reed, because it has to come to a single point. The mouthpiece of an oboe has to be very skinny: it's a consequence of the shape.

That's also why the orchestra tunes to the oboe, because it's shaped like a cone. There's no place in the oboe where you have two cylinders of the same diameter, one slipping tightly around the other, and you can make one longer and one shorter to tune it, as is the case with other woodwind instruments. You can't tune the oboe like other instruments.

There are some intimidating names for the mathematics—it involves something called spherical Bessel functions—but you can do it at the level of high-school physics. But anyway, the point is that the oboe has all of the integer multiples of the fundamental frequency. That is to say, the harmonics that come along with the lowest frequency oscillation are at twice that frequency, three times, four times, five times, and so on. But with the clarinet, when you go through the same song and dance, you only end up with the odd integers. The even-integer ones are missing. I found that very interesting.

I then took some freeware on a computer and programmed a sound generator with a bass note and a lot of the integer multiples of the fundamental frequency. Then I took a second program and took out all the even integer multiples.

I played one and I played the other, and I nearly fainted because there were no reeds, no bell at the bottom, no pitch holes, no fine choice of wood: it was just one computer with a bunch of frequencies in it and another one with half the frequencies missing, and you could tell which was clarinet-like and which was oboe-like. That really excited me.

However, the flute *also* has all integer multiples, but a flute sounds **nothing** like an oboe. The story was good for telling the difference between the oboe and the clarinet because they're otherwise quite similar and this is a huge piece of their difference, which you can hear just from the computer signal generator.

The flute has all the integers present too, but the question is, *What are their relative strengths?* The flute has a very pure tone and there are very few of them. You don't have higher and higher frequencies mixed in with any appreciable strength. So appreciable strength becomes an issue. What a good microphone and a good computer can pick up and identify as a tiny piece of the signal is different from what we hear.

Now there was also some heartbreak associated with all of this, but I think it's really the fault of the parents, not my students. I wanted to do a demonstration with the real instruments to see if people could tell which was which. I played them the introduction to *Peter and the Wolf* in which each animal has a solo instrument and a theme throughout the story.

HB: OK, but where does the heartbreak come into all of this?

DP: Three years, three classes, not one student knew where the music was from. They could tell which was the clarinet and which was the oboe—

HB: Ah—this is social commentary again. All of this was just leading up to the fact that you're very upset about them not being able to recognize *Peter and the Wolf*?

DP: Totally! These are multi-instrumentalists. These students are the first violin in some youth orchestra, and that sort of thing. It was shameful.

Questions for Discussion:

1. Why, exactly, do you think that David rejects Howard's analogy? What do you think he means, exactly, when he talks about the need to "make contact with music as music"?

2. Might there be a way to objectively quantify when two instruments—such as an oboe and a clarinet—are "relatively similar"?

VII. Transient Growth

Coupled, damped oscillators

DP: There is one more physics thing that we should touch on. I mentioned earlier how vibrating strings sometimes die down and then get bigger. I stumbled on a paper by Gabriel Weinreich, a professor of physics at the University of Michigan.

He'd written a paper about the piano and what makes the piano sound different from other earlier keyboard instruments. He had fallen into this late in his life as a physicist; and he's very well respected in the field of musical instrument acoustics, but he definitely has physicist tastes.

His question was, *How is the piano different than these other instruments?* It turns out there are many differences. There's one thing that he thought was really important, because the piano is both loud and has a long sustain of the notes—the notes linger longer, unless you damp them out. In fact, that's why you need a damper, unlike, say, a harpsichord.

I never thought of this but it's in the paper—you open up the piano and, for almost all of the keys, the hammer hits three strings. Almost every note is three strings. There's a small section of twos and a tiny section of ones, and the ones don't sound like music at all; they're very low.

His story was about the effect of hitting three strings tuned to the same note and letting them ring. It's a very important piece of physics.

Each string is what we call an oscillator. The three of them have the same frequency and they talk to each other because they're attached in the same place. There's a bridge, or terminus, to the strings, and one of them is going up and down, pushing up and down on that, and the next one feels that up and down, so they can talk.

What happens during this "talking"? It's a very important concept. We'll talk about a swing in a playground. You have a kid on a swing. You push the kid and you can get him going really high, even with a gentle push, as long as it's at the right time.

If you push at a steady pace, if it matches the natural pace of the swing, the swinging gets bigger and bigger. That's called resonance. That's a general feature when you have two things that have the same frequency. If you let them talk to each other, they can have a huge effect, even if the talking is very quiet and a little at a time, because it keeps happening.

If you push the swing just a little too soon, at first the swing will get going, but after a while you'll find that you're out of sync, and when the swing comes back towards you and you push it, you end up slowing it down instead of speeding it up. It's really important to be very close to the same frequency. So in the piano, you've got these strings having a very large effect on one another because they're tuned to the same note.

He begins his paper by saying, *"This is the subject of 'coupled, damped oscillators', which arises in many fields, but is generally poorly understood."* So I pursued the subject, and I wanted to apply it to my banjo strings because I was convinced that there was some connection. I was very interested in how the banjo strings talk to each other.

There are two strings that are adjacent to each other, connected to this bridge that likes to move a lot. They are each tuned to a different note, but part of those two different sounds includes the same frequency, which takes us back to the harmonics I was talking about earlier.

In fact, if I damp all the strings but one, and play it, it has a different sound than if I played that string without the rest damped. That's the string that I plucked talking to the other strings.

That's called coupling, and the *big* coupling, on the banjo, is the piece of the one string which has the same frequency in it as the fat string. Those are two different strings, but they talk to each other. But there are a lot of them that talk to each other. That's what was interesting to me.

HB: Was the idea that the sustain lasts longer as a result of this shared resonance, or what? What, exactly, is the claim?

DP: First of all, it's nine times louder if you have three of them, not three times as loud, because initially they're all pulling on the same bridge. It's not like each one is pulling separately on the bridge—that would be three times as loud—but the bridge is doing more because it's actually moving more.

The next item, which is called psychoacoustics, is the decay of the sound, which is not like a single oscillator. The thing you study in freshman physics is a mass on a spring and it's got some damping, and the amplitude gets smaller and smaller: the period stays the same but it just dies down.

But in this case it's not a single exponential. There are some long modes. We hear the loud at first, and then we hear something with a much longer half-life, which wasn't the first one, but first we hear loud and then we hear long. So that's different than a single exponential which has a single half-life. It's got these long components, which are combinations of the string, which only let the sound out a little at a time, because they hide it from the sound-making mechanism.

On the banjo, when the strings go up and down, they force the head to move, but when they go sideways they don't—they're still vibrating, but they don't force the head to move—so they're not making sound when they're going sideways.

Now a bit of history, which leads to the heartbreak—the could-have-been-famous issue.

Not only physicists take physics courses in college. Applied physicists, astrophysicists, astronomers, physical chemists, chemists in general, electrical and mechanical engineers—all these people take the basic physics syllabus, which includes the mass on the spring and adding the damping and seeing it decay exponentially. It also includes coupled pendulums, where one is doing something to the other, and they can move together or in opposite directions, or you can have one going which then gets the other one going, and so on, producing beats.

And then they do the *forced* oscillator: *If you come in from the outside and just keep pushing steadily, because you have a power source, how does the system respond?* It turns out that if you're close to one of its natural frequencies, it resonates, and if you're far away, it doesn't do much, and there's a formula for all that.

But they don't do the coupled, damped oscillators—the so-called "transience"—how, exactly, do coupled oscillators die down once you disturb them?

There's a reason they don't do it. The simplest problem would have two oscillators. And what you discover is that, if you damp these oscillators, there are special cases where it's really simple, but, in general, in the generic case, to solve the problem on the blackboard you need the solution to what's called a quartic polynomial.

In high school we learn the Babylonian formula for a quadratic. If you look it up in a book, you'll find that there's also the cubic—it was figured out in the 16th century. A student of the cubic guy figured out the quartic, and much later mathematicians proved that there's no generic solution for higher-order polynomials than the quartic.

The problem is that the solution to this fourth order polynomial, the quartic polynomial, can't really be conveniently written on a blackboard. You type it into your computer with appropriate math software and it spits one line that's four pages long. You can't look at it and know what's going on. As a result, people don't do that. That's why it's not in the syllabus. If you go to advanced textbooks, they're misleading because they say, "*Well, this is the problem, and it's just like the other one, but it's more complicated.*" They don't tell you anything interesting.

In many fields, however, this problem arose, and people had to figure out their own way to deal with it. But it was not part of their tool bag from elementary physics the way the spring, the pendulum, the falling ball—there are things that every physicist learns, but this is not one of them.

If you look carefully, however, you can see concrete examples of people identifying the relevance and importance of this.

One example involves Bruce Winstein, an experimentalist at the University of Chicago who worked on neutral kaons, which, for a long time, were this font of astounding information—that's the system where we first discovered that the fundamental laws are asymmetric in time.

Anyway, when Winstein gave a general lecture about his work, he came in with a system with two pendulums that he'd built in the machine shop, with a weak spring coupling them and an adjustable damper on one of them to show some of the phenomena as it applies to the neutral kaons.

The next example I found was from one of my colleagues who has since passed away, a mechanical engineer interested in earthquake safety. You've got a building, you kick it at the bottom with an earthquake, and you want to know what the building does. And what they found out from computer simulations is that, very often, there's a jolt, which causes the building to start swaying and, somewhere along the structure, the swaying gets big. Now, if it doesn't sway too much, it may die down, but that growth, if it goes on for too long, could break the building. Nowadays we call that transient growth. The idea is that the motion in some part gets bigger before it finally dies down.

HB: And this is reflected by the quartic equation?

DP: Well, if you could solve it—I didn't bother to do it—you would find it there. All he knew was that there were simple cases where this didn't happen at all. He could prove it. And engineers for a long time have known that, if you program a computer to just solve Newton's laws, to find the motion—cause we've just got springs and dampers—if you put it into a computer, it's got some wild solutions that are very sensitive to the parameter values.

The final heartbreak came about fifteen years ago, but first we have go back one hundred years to Reynolds of Reynolds number fame. The problem that he was interested in was the flow of water in a pipe. If the flow is reasonably slow it moves smoothly. Basically, the pipe and its intermolecular interactions make the water come to rest at the surface of the pipe and it's moving fastest at the middle. You can ask, *How fast does it move depending on where you are in the pipe?* That's something you can solve. The flow lines are straight along the pipe.

Now, if you speed it up enough, at some point, it goes turbulent. His genius was to understand that the point at which it becomes turbulent can be characterized by a simple, dimensionless number —the so-called Reynolds number—that involves the diameter of the pipe, how fast the stuff is going, and the viscosity of the liquid. So when you exceed this number the flow becomes turbulent.

The theorists then apply a technique which they'd used successfully in many cases—we don't understand, in terms of pencil and paper, turbulent flow, but there's a line of analysis we use to determine when something is unstable or not.

Suppose you have a heavy fluid on top of a light fluid just sitting there, then the heavy fluid should fall down. For example, if there was a bottle with water sitting on top of air, the water should fall down. But the fact is that the air has to get out of the way. So it's possible that if you very gently set up an apparatus like that, the water would just sit there—and the way a physicist would look at this is to imagine that the interface is not completely flat and smooth but rather has a little ripple in it. And now you ask, *If I start with a little ripple, will that get bigger or will it die away?* There's a lot of stuff to make things die away, like viscosity and surface tension. Anyway, that's the question: *If you put in a little bend to the surface, will it get larger or smaller?*

For a given situation, you can actually find the length at which it grows, and that's something you can see—they're called salt fingers in lakes. I learned this from teaching a course. Basically you have the salt water above the fresh water and there are spots, about the width of a finger, through which the salt likes to go down. The guy who discovered this, John Taylor, knew the equivalence principle, which means that the same physics applies to explosions. If you want to blow a light, less dense material through a heavy material—the heavy material has more inertia, the light stuff wants to go faster, how does it finger through? It turns out that it's also important to stellar structure. It's based upon the general idea of taking something that's stable and asking how it responds to little perturbations.

So we now take the pipe flow, where we know what it's doing when it's all moving straight and we ask, *What if it deviates slightly from just going straight? Will those deviations get larger or smaller?* And over the decades, with more and more mathematical precision, people showed that every conceivable little wiggle will actually get smaller. None of them grow. The conclusion was that pipe flow is stable to small perturbations. Therefore, it must be surface roughness even though Reynolds number seems to describe all the pipes.

The key to that problem was a rediscovery of what Gabriel Weinreich knew for piano strings and what Murray Gell-Mann knew for neutral kaons. This is something that math people understand. There are frequencies, which are eigenvalues, and there are motions, which are eigenvectors. And the eigenvectors are not orthogonal in this problem.

Now going back to the coupled, damped oscillators: in general, if you only have the coupling, you can find some variables in which it decouples. It's a bit complicated but there are motions that occur totally independently of the other motion. Once you have the damping, it generically mixes those up, and there's no motion that's separate from the other motions. The eigenvectors are not orthogonal.

It turns out that this is a very generic feature of many systems, except for the ones that we teach in undergraduate and even *graduate* physics.

Many months ago, I asked my physics colleagues and not one of them had thought about it or heard about it. They had some interesting things to say, some suggestions of things to pursue but couldn't tell me exactly what happened. By the time I figured out what happened, I had asked around and was directed to a guy in applied math and he said, *"Oh, yeah, that's really embarrassing. We figured that out 15 years ago in applied math and in fluid mechanics."* It's called transient growth. It's now in chemical physics and I'm sure there are even economists who are interested in it.

So what are we talking about? Well, even if you have one string, it can go up and down and it can go sideways and, to a first approximation, they don't do anything to each other, but they talk to each other a little. So, if you started doing one, after a while it's typically doing another. Furthermore, by the time you're done, instead of having two with exactly the same frequency, you always end up with two different frequencies. You can't get rid of the beats. That throbbing is the presence of two different frequencies.

Then you add the damping, and it mixes the things that have a fixed frequency. So now the question is, *What happens?* Yes, there are particular motions which have a given frequency and a given damping time, but when you add them, now there are different things that come out—I don't exactly know how to say it without writing down formulas.

But if you remember back to high school arithmetic, when you solve the quadratic equation, there's a square root and the thing under the square root can be positive or negative. When it's negative it's trouble, because that's not a real number.

In short, then, we learn something about the solutions that are different depending on the sign of that thing in there. With two oscillators that are coupled, even if the coupling and the damping are both weak—which should be a simple problem that we can visualize —depending on the sign of what goes under a square root, the nature of the overall behaviour is radically different.

I should also say that the key from Gabbie Weinreich was to realize that we're not going to write the quartic formula on the board but, if you start out with two oscillators that have nearly the same frequency, we can solve the problem approximately, which is what he did in equations.

I thought it was worth rewriting it for no other reason than all the people who cite his paper, which is very famous in the acoustics of musical instruments, mostly don't understand what he's talking about. They roughly know the result, but he does it in a rather sophisticated way—his physics was always more sophisticated than mine.

Anyway, I realized that there was a way that it could be done on the blackboard in the sophomore syllabus.

Rayleigh wrote about the problem in the late 19th century— he was another great physicist who spent his life doing sound and acoustics and writing about it and having insights. He recognized the problem as an unsolvable one except in some extreme conditions, but he doesn't give you insights about the general problem and that's what ultimately came from applied math and fluid mechanics, that the generic problem has these non-orthogonal eigenvectors.

Questions for Discussion:

1. *Are you surprised to discover that a generic feature of many physical systems has long been ignored by many physicists? To what extent do you think that this might be explained by an ever-increasing amount of academic specialization?*

2. *In what ways does the reinterpretation of a specific phenomenon within a particular mathematical framework (such as those involving eigenvectors, as David mentions in this chapter) result in a much deeper understanding of nature? How can this be used to distinguish between the relevance and applicability of different mathematical frameworks?*

VIII. The Working Physicist

Ruminations from the front lines

HB: This story of transient growth you've just told raises a few interesting sociological points, in my view.

At the risk of simplifying things too much, there seem to be two types of theoretical physicists when it comes to a general operating philosophy. There are those who are intent on discovering "the one big fundamental equation" that, in some deep, abstract way, underpins everything, while there are others who prefer to focus on exploring what's around them, asking themselves if they truly understand the things that we think we do.

It seems to me that this second approach resonates strongly with your style. Earlier you briefly alluded to your work in asymptotic freedom for which you won the Nobel Prize, once more describing it as a case of you asking something to the effect of, *What's actually happening here?* There's something going on in the world that we don't understand and let's see if we can do our best to understand it.

Most people would probably imagine that winning a Nobel Prize in theoretical physics would require the highest possible level of abstraction, quite removed from this essential, down-to-earth perspective of trying to puzzle out what the heck is going on in some specific instance. And yet, you seem to bring that essential, particular curiosity to all of your work. Is that an accurate description of your situation, you think?

DP: Totally. That's right on. I've known outstanding physicists of both types. There are the ones who don't even want to think about something unless they see its possible connection to the most important problem. One of my good friends likes to characterize it as that you want to do something that will be in the textbooks for the next generation, something really big and important. Okay, sure, but it's a question of how you get there, and even if that really *is* your goal. Because, as I described earlier, it's really just about the joy of figuring something out.

Why do you do physics? For some people, it's a desire to get to the bottom of things, and I'm pretty sceptical of that: we don't get to the bottom of things, but we get somewhere.

I've been very conservative in my physics, and my interest in the banjo is similar—I mean, I'm not trying to invent new kinds of instruments. There are things I think we should be able to understand.

There are a couple of examples in current theoretical physics— what's called particle physics now, which has a very broad definition—which I find very gratifying, because in the last ten or fifteen years, I've thought that these really are the most interesting questions, and yet they weren't at the forefront for some time. I had just run out of the kind of intensity that's required to do particle theory at the professional, frontier level. For me that was a 24 hour a day, 7 days a week activity.

Anyway, the two examples are, first, Hawking radiation—the radiation from black holes. It was a great discovery by Hawking and Bekenstein, but to me there were always questions you could ask; and I felt that if you understand physics, you should be able to view it from a different perspective and get to the same answer.

But for years the general response was always something like, "*No, that's too hard. **This** is the way we understand it. **This** is the way to do it.*" But just in the last five years or so there's been a resurgence in interest, a realization that there's something ***profound*** that we don't understand there.

Something like it must be true. At various times in the past there were bottles of champagne bet and encyclopedias going back and forth, and people convincing other people of this and that, and things apparently settled—but *not really*, I think. I want to know what is really happening there. What's ***really*** going on?

HB: So, this is the question of information loss in black holes? What exactly are we talking about here?

DP: The firewall business, the information stuff, various perspectives said that it can't be the way we understand—I mean, a lot of it has to do with quantum mechanics versus general relativity as a classical theory.

Kip Thorne, who was a student of John Wheeler, gave a talk in remembrance of Wheeler. Wheeler invented the name black hole—he's responsible for a lot of good names—even though the concept goes back quite a bit. And there was this issue about what comes out of them. His students convinced him that *nothing* comes out, because he had taught them general relativity; and according to general relativity, nothing comes out.

It really **bothered** him, and they finally just beat him into submission: they must have just outnumbered him or something. In the end, there is something to what he said: *We're not making* **contact** *with quantum mechanics.*

And not just by settling the fundamental issues like string theory is trying to do—though some people say that's necessary to resolve these issues—but just what's happening at or near the horizon because of the information issues, because of features of what quantum mechanics tells you how mechanics really works. It doesn't work classically for the particles, even with some simple-minded version of gravity

Anyway, there's been a huge resurgence in that.

The other thing that is a major, active field is that, when you do calculations in what we call non-abelian gauge theories, which are the equations for the standard model—the strong force, the weak force, the electromagnetic force all have this same mathematical structure —you do complicated calculations and then the answers come out in an embarrassingly simple form.

So you've got a notebook with tons of pages all covered in formulas—we have an algorithm for generating this stuff—and people invented huge advances in computer mathematics, which were prompted by the desire to do these kinds of calculations (such as Stephen Wolfram's *Mathematica*).

And at the end it collapses to some ***very simple form***, without us really knowing why. So that has also become a major field now, both to get to the simple answer more efficiently so as to best analyze data from the biggest accelerator—the LHC at CERN—but there's also a fundamental question about whether we're thinking about it wrong. Many people think that we are. I've long been concerned about that too, but the mathematics involved in this discussion is beyond me at the moment.

This is a small, specialized field and they're totally willing to give up all fundamental principles and look for some compact set of principles which will generate the stuff that we came to believe is true. Now, what we came to believe in terms of quantum mechanics, relativity, and then, ultimately, the models of particle physics, is very awkward and backwards and historical: we had electrons and photons and we had insights about very detailed features and very small effects which can be computed with huge accuracy—the thirteen significant figures of the electron's magnetic strength, for example.

But then we got to quarks, and we have a mathematics that has a very similar structure, and yet you're not supposed to be able to have a single quark; and there are features of that interaction that are really different and weird.

We got into it by thinking that the quark is like the electron, but *is* it? Well, we know it isn't. And the forces that hold the quarks together are kind of like photons, but they're not, because we can see and do things with photons, and we don't do anything with those guys.

They don't really have anything to go on except for the final answer. It's kind of weird. They're not doubting that we have some theory which describes nature to a great extent.

And then they do something which theoretical physicists often do: they say, "*Well, that one's too hard. I've got this simpler one, and I'm going to work on that one.*" So you've got a number of people working on this one, which is obviously not the theory that describes the particles we know, but it's close. And now you can ask, "*Well, can we get some insight about its fundamentals?*" It's not a crazy approach.

The imagination here is that there will be lessons learned. For some reason, there are answers that are simpler than they should be, and there should be lessons learned. And then you come back to the main problem.

I remember teaching freshman physics and was teaching the pendulum or whatever, and one student asked me, "*When are we going to get to the **real** stuff, where you don't ignore friction and all of that?*"

Well, this is one of the treasures that Galileo gave us—a profound gift to the human intellect: *Let's ignore friction for the time being and we'll come back to it, but let's put it back in terms of the concepts we have developed in a world in which we had ignored it*, because if you don't ignore it in the beginning—well, just think of Aristotle: it sits on the floor because "that's where it wants to go", it does "what it wants to do".

HB: You mentioned particle physics, and the fact that you used to be thinking about it 24/7. Obviously these fields move in all sorts of directions and one needs a tremendous amount of intensity and excitement and so forth to make contributions.

Equally obviously, you're very excited about doing your current work, but do you ever feel torn sometimes—not just because of the appeal of fascinating outstanding issues in particle physics, but also in completely different areas of physics entirely, because there are so many interesting problems to work on? You go to a seminar and think, *I'd like to know more about that, but I have to manage my time because if I'm spending all my time learning all of these other things, then I won't be able to advance with my own research*. Do you ever have any of those sorts of quandaries?

DP: Well, I know myself well enough to know that I can't do it. You have to imagine that you have a chance of succeeding. It's hard.

I've got a few years between then and now, and there certainly were times when I made a stab and some of them I thought were respectable—not world-shaking, perhaps, but respectable. It had the same personality characteristic: I sunk my teeth into some thing that I thought I should understand.

It's now been some years, but you might remember the first guys who succeeded in making what are called Bose-condensates out of gases. That was a hot experimental field before they succeeded.

There was a guy here at Caltech who had a student who was trying to do it, and I got quite interested in that. There were new things to learn about what is now called nanoscale material, the solid-state physics.

I was intrigued by one simple fact, which is kind of marvellous: it's a feature of the physics of the very small. Heat in a solid is represented by vibrations—we call them phonons—the quanta of vibrations, which have wavelengths. When it gets cold, those wavelengths get longer and longer. You can make something small enough, and it's got billions and billions of atoms—as Carl Sagan would say—and it can be small enough and cold enough such that the wavelength of the relevant heat quanta is much bigger than the thing.

So it's essentially at zero temperature, because it has no thermal motion. It's in contact with things that have a temperature, and it's in equilibrium with it as a temperature, but there's no motion there. So you're now at zero temperature, and you can then tease apart—and this became a huge field of condensed matter physics—the thermal physics from the quantum physics. We used to have really simplified pictures of both of them and there are some heroes in the history of quantum mechanics who've had insights about solids which have essential, quantum features. But we've made huge progress. I did a little bit of work there.

There were questions there that I thought we should be able to answer. Let's put it that way. It wasn't hugely speculative. There was a way to carve out some very straightforward, Politzer-like, question that would have an answer; and some of it I was able to answer. That made me feel good.

Since then, the issue of the black hole horizons and the information and the radiation, or the other one that I mentioned about gauge theories being simpler than we think and reformulating the whole thing—I don't have the strength to work on those. It's like being a graduate student again. You can't do it part-time, or feel like you've gone into a room to get something and then you don't know why you're there. That's hard if you're doing theoretical physics. But everybody's different. You spoke to Freeman Dyson recently (*Pushing the Boundaries*). He's still going strong. ·

HB: He's a remarkable fellow.

DP: Yes, he is.

Questions for Discussion:

1. Are some scientific disciplines inherently better suited to younger researchers than others? If so, why do you think that is?

2. To what extent does the rapidly expanding mathematical requirements in a field like theoretical physics decrease the likelihood of being able to make a genuinely interdisciplinary breakthrough?

3. Is there a difference between "the joy of understanding" something and "the joy of discovering" something? If so, what is it, exactly?

4. Do you agree or disagree with David when he says, "We don't get to the bottom of things, but we get somewhere"?

IX. The Journey Continues

Joys, frustrations, and the banjo brotherhood

HB: Well, I'm almost done—you've been very generous with your time, thanks very much. I'll ask a question I've asked many others: if I were God and I could answer any research question you might have, what would you ask me?

DP: That's not the point. The point is the process. It's like the Toyota commercial: *"Life is the journey."* I think it's the journey that matters, not the answer.

HB: But there's nothing keeping you up at night? It can be a really small question. It doesn't have to be the meaning of everything.

DP: Earlier I expressed my frustration about the gap between what we can measure, or think we can measure, and what we extract for understanding. I told you that I loved the story about the oboe and the clarinet, and I thought that by trying to understand how the tailpiece worked—that was the simple question—my proposal was that *that's* what makes banjos different from everything else.

Now, I don't actually know if that's true, and I don't know quite how you would answer it. I've spoken to professional acousticians of musical instruments and they have their own ideas about what it would mean, and I'm not sure about that. So I'm a bit confused. So I'm guessing that the Almighty would tell me, *"That's there, but really it's a hint that you should look more closely at x."*

There are other factors that I'm going to get to that might be characteristic—as I said, I believe that a drumhead, this kind of bridge, and strings, is what all these banjo-like instruments have in common.

What I like to point out to people who talk to me about different types of strings, or weights, or what have you, as being responsible for the characteristic sound of he banjo, is that Stephen Foster, America's first great professional songwriter, in many songs refers to "ring the banjo."

His banjos had no metal parts. But a banjo still rang. Banjos rang long before there was any metal anywhere near them. They had the gut strings and the skin with the thin rims, or they were gourd banjos, yet they rang. So what is it about them that makes them ring?

There was an interview with Keith Richards when he was asked if he had ever played the banjo. And he said no, because it was just too mysterious to him: it just rings. So the banjo rings, and I'm not even sure what that means. What should I be looking for?

When you read a description of John Chowning's work he said, "*Well, I did acoustic frequency modulation and it rang.*" You can clearly hear it. Is that what you're hearing? There are other people who say, "*No, you're hearing this combination of such and such.*"

And somebody else points out, "*Well, there isn't a mathematical distinction. If the strings weren't damped and it was all steady, the thing going up and down would not introduce new frequencies; it would just change the relative heights.*"

Is that what ringing is? The electrical engineers know about frequency modulation because that's FM radio, and they talk about side bands and bandwidth and all kinds of things. We know about that stuff. But the situation here is a little different.

In my characteristic way, my next project is totally different and much more modest. It concerns one particular style of banjo. It sounds different. You can hear it. Some people like it, but most don't prefer it to alternatives. I've built ten variants of something clamped on the back that mucks around with things, and I record them and note the differences in sound.

Here I have a machine that's shaking it, so it's not related to any movements I might make, and I stick the microphone up close and see if I can connect the geometry of what's inside to what I hear.

It's not working yet, but I like that frustration because sometimes it gets better. I have to remind myself of that. At the time, I'm very frustrated. We're a very temperamental bunch. When you finally do something, then it's done—and then you're depressed because there's no sense of accomplishment. I'm sure there are some theoretical physicists who feel some great sense of accomplishment with what they've done.

But, by and large, the thing that you do becomes public property; and when you meet other people, they know your name and they smile. And they always want to know what's new: what have you done since? Because they already know about the other stuff. So the rewards are few and far between.

HB: Well, there's the personal, ongoing reward of doing it, right?

DP: Well, it's frustrating. As a theorist your garbage can is full of crumpled paper. So having a piece of hardware to work with gives a nice aspect to things, I must say.

As you can see, in this model I also developed a special piece to put behind the banjo for testing so that I could have a solid back that wasn't my belly that gave me an objective, reproducible gap. Acoustical engineers said that closed cell foam is good for absorbing, and cork is good for reflecting, so I incorporated that. I added a Hawaiian shirt too, because that's what middle-aged guys wear. So now I have to try that on the back of the banjo and we'll see how it works.

I like it that I can pick up the banjo and play it, and not only say that I'm working, but know that I'm working. That's a nice feeling.

HB: That's perfect. Anything else you want to add?

DP: Well, when I was active in particle physics and giving talks and travelling around the world, there was this phenomenon—my advisor referred to them in Italian because he had, many times, gone in the summer to give physics lectures in Italy, and he referred to them as *i fratelli fisici*, the physics brothers. The point is that you could go anywhere in the world and there were people in your field whom you had never met in your life, but they meet you at the train at 6 a.m. I went once, for instance, from Moscow to Leningrad and there was a guy meeting me at the train, and this guy knows my work, and I know his work, and we're like old pals. And we start talking about all kinds of things.

That was back when I used to travel by air, by train. Now I travel electronically. I've mentioned Gabbie Weinreich a few times already. Jim Woodhouse was very helpful to me—he's an editor of one of the two most respected journals of acoustics. Dan Levitin was a buddy of someone here. He passed through, and I told him I used his book and I had questions. I told him that one of my students did a project and they discovered some things, and he replied that he had a graduate student who did that and so on—so there's a real connection to people.

This also means connections to luthiers, such as Greg Deering, I mentioned Jens Kruger, earlier, a world-class banjo player. He's a consultant for Deering Banjos, and I've had the chance to talk to him at great length about hardware. Many musicians have little interest in actual mechanics, but he's different, and he has a wonderful ear. He can tell you when there are lousy strings, or bridges. I don't know what he's listening to, but he can hear it; and he's had many ideas over the years for improvements to the instruments.

And then there's Greg Deering himself. Even though he recently passed the 100,000 mark and can't meet demand, he's still looking to expand. His biggest retailer is in Great Britain, and he's increasingly looking globally. There are still places, apparently, where they don't have Deering banjos. But as an entry-level instrument of various styles they're really excellent. It's been great working with him as well.

HB: You mentioned the physics brotherhood, but this seems to be another sort of brotherhood.

DP: Absolutely. While most don't, there are a number of banjo players who do a lot of fiddling with the hardware or who make banjos themselves—sometimes really marvellous ones. Somebody posted this on one of the electronic bulletin boards: *"How many of you are engineers?"* and there was a long list. It's not that most banjo players are engineers, but, in fact, it's a non-trivial overlap. Not so many physicists, though. The ones I knew in the academic world were mostly biologists, actually.

HB: Well, maybe you've started a trend. Maybe there will be thousands of physicists who will start playing the banjo.

I've enjoyed this very much, David. Thank you.

DP: Yes, it was a pleasure.

Questions for Discussion:

1. Are you surprised at the idea that banjos "rang" before they had any metal in them?

2. Is there a risk of diminishing our enjoyment of musical experience by trying to develop a comprehensive scientific explanation for what we are hearing?

The Problems of Physics, Reconsidered

A conversation with Tony Leggett

Introduction

The Gentleman Laureate

The Nobel Prize has always vaguely irritated me. The idea that one's entire research career might somehow be neatly defined by what a bunch of Swedes happen to find noteworthy has long struck me as arbitrary at best and, in my darker moments, a sad commentary on our collective need for self-affirmation.

Richard Feynman, typically, summed it up best when asked if his work on quantum electrodynamics fully merited being awarded the Prize: "*I don't know anything about the Nobel Prize and what's worth what...I've already got the prize. The prize is the pleasure of finding the thing out, the kick of the discovery, the observation that other people use it. Those are the real things. The honours are unreal.*"

On the other hand, it's clear that these sorts of major prizes and awards have their uses. Life without the annual Nobel announcements, for example, would mean that the mainstream media would pay even less attention than usual to scientific discoveries, literary accomplishments or the enlightened few who are advancing the cause of global peace.

The Nobel Prize does one other very useful thing too: it provides a sort of "academic bulletproofing" for those who might wish to indulge in more general musings on the future of their field, speculations which are great fun for the rest of us who might be unwilling or unable to follow technical arguments in detail.

Along with two others—Alexei Abrikosov and Vitaly Ginsburg—Tony Leggett won the Nobel Prize for physics in 2003 for "pioneering contributions to the theory of superconductors and superfluids" and

is universally appreciated as that very rare bird indeed: an unequivocally brilliant and unreservedly decent person. In a world filled to the
brim with tedious academic posturing and self-proclaimed geniuses,
Leggett has earned the reputation, as a friend of mine once pithily
expressed it, as the one Nobel Laureate that people would most
enthusiastically invite over to dinner.

He is also, in his own quiet and unpretentious way, one of physics'
most penetrating and thoughtful popularizers. In 1987, one year
earlier than Steven Hawking's *A Brief History of Time* and more than
a decade before Brian Greene's *The Elegant Universe*, Tony penned
The Problems of Physics, an insightful, plain-speaking itemization of
the physics landscape according to four basic categories: the very
small (particle physics), the very large (cosmology), the very complex
(condensed matter physics) and the very unclear (foundations of
quantum theory).

The book is a delightfully written summary that is surprisingly
relevant today, notwithstanding all of our modern advances. Likely
for these reasons—not to mention that whole Nobel Prize business—
Oxford University Press elected to re-issue it in 2006.

The last category of *The Problems of Physics* is a particularly intriguing
one. Back in 1987, discussions of foundational aspects of quantum
mechanics were regarded by the vast majority of working physicists
as roughly on par with astrology or alchemy in terms of respectability.
Tony, however, 16 years before his bulletproofing Nobel was awarded,
clearly felt no hesitation to boldly stroll into the quantum labyrinth.

Indeed, he has spent a significant percentage of his professional
scientific life quietly probing the foundations of quantum mechanics, determined to find where the theory will break down. On the
surface, this might not sound terribly surprising. After all, quantum
mechanics has long been recognized as a framework fraught with a
bevy of conceptual and logical difficulties that has tormented some
of the best scientific minds the world has ever known, including

luminaries such as Albert Einstein and Erwin Schrödinger who did so much to develop the theory in the first place.

But precisely for this reason, modern physics has distanced itself from efforts to penetrate the mystery that is quantum theory, summarily branding all of that business as "intractable philosophy", while anxiously moving on to attacking problems it might actually solve (such as superfluidity). After all, however it might trouble us conceptually, it can't be denied that quantum mechanics works like a charm.

Once again here's Richard Feynman, to set the tone: *"Do not keep saying to yourself, if you can possibly avoid it, **'But how can it be like that?'** because you will go down the drain into a blind alley from which nobody has yet escaped. **Nobody** knows how it can be like that."*

Tony Leggett, however, in his ever so polite, understated way stubbornly refuses to give in:

> *"What really worries me is Schrödinger's cat. The formulas of quantum mechanics haven't changed a whit as we go from the description it gives of the photon to the description it gives of the cat. If we refuse to make a particular interpretation at the microscopic level, then we have no business reintroducing that interpretation at the macroscopic level.*

> *"The evidence for this statement is there at the microscopic level in the form of interference patterns, but everyone agrees that, at the level of the cat, it has gone away. But does the fact that the evidence against a particular interpretation of the formalism has gone away by the time we get to the cat mean that we can freely reintroduce that interpretation? I say, 'no'."*

What separates Leggett from many of the others concerned about the foundations of quantum theory—and incidentally unites him with the likes of Feynman—is his insistence on the importance of experiment to probe our theoretical constructs:

"Perhaps I'm just ultraconservative, but my attitude has always been that physics is an experimental subject; you don't want to push your theoretical speculations too far beyond what we can currently access experimentally. I suppose that's why I've always stayed within the confines of condensed matter physics, or things somewhat related to condensed matter physics, because I like the fact that, if I have an idea, there is some hope that my experimental colleagues will be able to test it during my lifetime."

Three decades later, the physics world has caught up to Professor Leggett. The rise of quantum information theory, quantum computing and quantum cryptography has breathed new life into the business of probing the limits of quantum theory.

"When I first started thinking seriously about this, way back around 1980, I quite seriously hoped that when you got to the level of the so-called 'flux qubit'—where the two states we're talking about are different in the behaviour of something like, say, ten billion electrons—by that time something else might have happened. Right now, it looks as if quantum mechanics is working fine at that level."

When I pushed him to speculate on what physicists will believe 50 years from now, he had this to say:

"In 50 years, I think there will have been a major revolution in cosmology and I think there's a small but non-zero chance that we will have pushed quantum mechanics in the direction of the macroscopic world to the point where it will fail and break down."

Not much cause for enthusiasm, then, for quantum philosophers anxious to see some revolutionary developments in the short term. But what if we take a still longer view? *Will quantum mechanics definitely break down at some point?* I pressed him.

"Yes", he responded, firmly and unhesitatingly.

Tony Leggett might seem like a kindly English grandfather, but the scientific will that drives him is as hard as steel.

The Conversation

I. Back to the Future

Setting the stage

HB: Do you remember what the general reaction was to your popular book, *The Problems of Physics*, when it was first published in 1987?

TL: I got a lot of interest from the sort of people I'd hoped to interest: people who are not professional physicists but have some interest in physics and moreover are prepared to do a certain amount of work on the subject. I got some quite favourable reactions.

Some of my colleagues liked it too. One or two, I think, took objection to one or two statements I had made in it but not many.

HB: But you were very careful. You had these phrases—I can't recall them off the top of my head—to the effect of: *"So far what I've done is orthodox, now I'm going to be much more speculative. If you have a problem with this, if you are a professional physicist, then you should be aware that I have written other things..."*

TL: Yes. I tried to guard my back as best as I could. There were some who disagreed, but they didn't take umbrage; they just wrote that they simply didn't agree with what I had said.

HB: There wasn't much in the way of popularization in 1987, was there?

TL: Actually, it was just at a time when a number of popular books, particularly on the foundations of quantum mechanics and its interpretational problems, were beginning to come out.

I think this was not necessarily the case for other areas like, say, cosmology, or problems relating to the arrow of time or anything like that. Even today I suspect that there are not many popular books in those areas. Nowadays, of course, one has a real flood of popular books on quantum mechanics.

HB: A veritable tsunami, as it were.

TL: Yes, indeed.

HB: I reread *The Problems of Physics* recently, the version that was reissued by Oxford University Press in 2006. I loved your categorizations: the very small, the very large, the very complex, and the very unclear.

I think it would be fun to go back and do a sort of follow-up analysis from our contemporary perspective using those same categories, assessing how much progress we've made and what mysteries remain or have come on the scene in the meantime.

TL: OK.

Questions for Discussion:

1. Why do you think that there are so many more popular books on the interpretation of quantum mechanics than other subjects in physics?

2. Is writing a popular science book more accepted today than it was 30 years ago? If so, why do you think that is? Are there any disadvantages associated with this change in attitude?

II. The Very Small

Much the same

TL: If one starts with the very small, then—and this is not an area that I've followed in great detail over the last 25 years since the book was written—I would say that the standard model of particle physics does seem to have worked rather well, by and large. As far as I know, there have been no huge pieces of evidence that have claimed that the whole thing is wrong. Of course, recently there have been these experiments that have discovered something that looks and smells pretty much like a Higgs boson, which is a nice, central piece of the jigsaw that does seem to be fitting in rather well.

HB: You say "looks and smells" but if I'm someone who doesn't know anything about this I may think to myself: *Well, Professor Leggett, you seem mostly convinced that this is really the Higgs boson, but are you really convinced, or are you waiting on mutual corroboration from enough people? Is there any lingering doubt in your mind? Or are you just saying that you don't know enough about this area?*

TL: I think I'm saying that I don't know enough about this area. To really make a proper assessment of—and I'm not sure if you can rigorously define this concept—the degree of probability that the Higgs boson has been found, then you need to go into the nitty-gritty of both the theoretical predictions concerning it and the details of the experiment. And I certainly can't. I'm basically basing my views on the opinions of my colleagues, and my impression is that it seems like a pretty good candidate.

HB: Sure. I should say that opening up *The Problems of Physics* was an interesting experience because I went into it thinking, *Well, 1987 was a long time ago. That was before dark energy, string theory was a very different thing than it is now,* and so forth. I suppose I was somehow naively expecting that I was going to look at this and find it to be completely out of date.

But it wasn't that way at all, of course, which really brought home to me how little our understanding of this area has changed in the intervening time. You mentioned supersymmetry, for example, which was something that people were talking about in 1987. And they're still talking about it now.

TL: Yes. By and large, at least in so far as I have any feeling for it, the picture of particle physics has not changed that much.

HB: I would think that there are probably more people who are talking about the idea of "physics beyond the standard model" now, than there were at that time. But perhaps not.

This is all speculative, of course, granted that this is not your area of expertise; but obviously you're aware of many developments. Is there anything you have a gut feeling about—a speculative gut feeling about anything in particular related to particle physics?

TL: To be honest, not really. It seems to me, for example, that the problem of reconciling the existing structure of quantum field theory with what we believe about gravity is still, essentially, as severe as it was 25 years ago.

Questions for Discussion:

1. To what extent does the brief discussion of the Higgs boson in this chapter demonstrate the extent to which professional scientists need to trust the opinions and valuations of their colleagues?

2. Is the fact that our understanding of particle physics has changed so little in the past few decades evidence of the strength of our current knowledge or the lack of sufficiently probing experiments? Or, somehow, both?

III. The Very Large

Cosmology

HB: Cosmology, on the other hand, has changed enormously since 1987.

TL: It certainly has. Obviously, I think, the most spectacular development has been the apparent evidence that the expansion of the universe has actually *accelerated*, contrary to what had been expected.

My own personal feeling—again I'm speaking as a complete outsider who is just trying to get a qualitative sense of what's going on—is that ideas like dark energy and so forth are really in some sense Band-Aids that people stick on what may in the end be a much more serious problem.

HB: I wanted to do the speculative part towards the end of our conversation, but I would like to make a lateral move to speculation right now because you threw that out there.

Granted that you're speculating, and that you have no hard evidence for this nor a hard theoretical construct, nonetheless your gut feeling is that this is a Band-Aid. Tell me a little bit more about that because that's fascinating. What sort of things are in your mind when you say that?

TL: I think in some sense these ideas concerning dark energy and so forth were formulated within the current paradigm of cosmology.

I have this gut feeling that we are in what Thomas Kuhn called a "pre-revolutionary stage"—things are going, in some sense, so wrong that we're going to have some kind of recognizable scientific revolution in this field, at the end of which it will turn out that not only have we been giving the wrong answers but we've been asking the wrong questions.

I really can't put my finger on it any more than that and it's no more than a gut feeling.

HB: Of course you can't put your finger on it, otherwise you'd presumably be a lot happier. But let me see if I can explore your discomfort a little bit more.

Do you suspect that this is tied to some foundational issues in quantum theory at some level? Do you think that these things might be linked, or is your sense that this is something just completely different?

TL: My guess would be that it is not linked to basic issues in quantum theory because, quite frankly, the whole subject—I'm going to offend a lot of my colleagues at this point—my whole attitude to the field of so-called "quantum cosmology" is that I believe it's simply not a subject, because I think we simply have no sufficient evidence to believe that quantum mechanics applies on the scale of the whole universe.

In some sense it's the default option, but we've had a lot of default options in the past and they've turned out to be overthrown at some stage or another. My feeling is that the problems of cosmology don't have much to do with the problems in the foundations of quantum mechanics.

HB: In conversation with you before we began filming, you mentioned the word "grandiose". And as you were talking just now, I thought that it's hard to imagine something more grandiose than the wave function of the universe.

TL: It's true.

HB: In fact it might well be the acme of grandiosity. That seems to be somewhat along the lines of what you're saying with respect to quantum cosmology.

TL: Yes. Perhaps I'm just ultraconservative but my attitude has always been that physics is an experimental subject and you don't want to push your theoretical speculations too far beyond what we can currently access experimentally.

I suppose that's why I've always stayed within the confines of condensed matter physics, or things somewhat related to condensed matter physics, because I like the fact that if I have an idea there is some hope that my experimental colleagues will be able to test it during my lifetime.

HB: That doesn't seem so over the top to me: for a physicist to expect testing of a hypothesis within one's lifetime.

So, let me try to summarize a little bit in terms of where we are. In terms of the very small—particle physics—our contemporary version of *The Problems of Physics* is that the standard model seems to be holding up reasonably well. We seem to have found the Higgs, a mechanism which was first proposed in the 60s.

TL: The early 60s. That's right, yes.

HB: So that's been around for a long time, and now, finally, it appears that there is considerable evidence for it. There are people who specu-late beyond the standard model of physics; maybe there's stuff out there, maybe there isn't, but either way we'll find out later on. So far there have been no revolutions, no hugely unexpected developments.

In cosmology, on the other hand, the extremely unexpected development was this discovery of the acceleration of the universe. And perhaps it's worth emphasizing for a moment that this is not an expansion of the universe per se—for a long time people have been aware that it's expanding—but rather it is the notion that the universe is *speeding up* in its expansion, or accelerating, as a result of so-called dark energy.

Regarding dark matter, you talked about that as a real problem back when you wrote *The Problems of Physics*. Do you think that is something that more people are paying attention to now?

TL: I suspect that even at that time a fair amount of attention was being paid to it. So, again, I think the situation has not changed qualitatively in that respect.

HB: In terms of your gut feeling that dark energy is somehow symptomatic of us moving from one paradigm to another, a consequence of looking at things through a pre-revolutionary lens, does your gut feeling tell you that dark matter is part of that, or is that something completely different?

TL: I'm not sure. I suspect it's something different, but again I don't have enough detailed knowledge to know how to answer that. But I suspect it's a different issue, actually.

HB: One other thing on the astrophysical front: you'd mentioned that people had started to detect supermassive black holes—or words to that effect—very, very large black holes at the centre of our galaxy, in 1987. I was completely unaware of that. I didn't know that was going on in 1987. I thought that was more of a 90s thing.

TL: Actually, I think what I said in the 1987 version was that there was speculation that there *might* be such a thing. I think now people are a lot more confident.

HB: Your summary was quite prescient, then, it seems.

TL: Maybe, yes.

HB: So that's another thing that seems to have changed in terms of our level of confidence now: astronomers now seem to be quite confident that most galaxies have these supermassive black holes.

Questions for Discussion:

1. Do you think that if scientists were more keenly aware of the history of science they would be in a better position to identify when they might be in a "pre-revolutionary" phase?

2. What do you think Tony means, exactly, when he talks about "the default option"?

IV. A Glassy Digression
The perils of affirming the consequent

TL: I should say that in many ways I'm a sort of perennial sceptic. I can't help wondering if the situation is such that we essentially have only one theory, basically Einstein's general relativity, and there are no serious competitors at the moment. Therefore, people have this automatic tendency to fit their observations to this one theory. I've seen this happen numerous times in condensed matter physics. I wonder if astrophysicists are not quite as immune to this as they may think.

I'm a terrible Popperian in the sense that I take very seriously the necessity to avoid what logicians call "the fallacy of affirming the consequent". Basically the fallacy is: *theory T predicts experimental consequent E; we see E; therefore T is correct.* That is a logical fallacy.

Now, of course, this fallacy is committed everyday in the pages of *Physical Review*. And the question is, *why?* And why don't people get worried about it?

Of course, the usual answer is that the projection is so striking and in some cases so counterintuitive that it seems very unlikely that any theory other than T would have predicted it. Therefore, we can affirm the consequent. But you are always making that unspoken assumption. So the question is, *How confident are you that no alternative theory could predict this?*

I have a particular example of this, which is a bit of a bee in my bonnet, that I've thought about for the last 25-30 years actually: the behaviour of glasses—just ordinary window glass, that kind of stuff—at low temperatures. It's quite puzzling, quite mysterious.

What is really mysterious is the following. I invite you to give me a black box containing some kind of material. The material has to be solid below one degree Kelvin, but that basically includes everything except helium as far as we know, so that's not a problem. And all I ask is that you ensure it is not crystalline and not metallic. And I will make a fair bet that I can predict the dimensionless absorption of ultrasound waves of a certain frequency in that material, below one degree, to within—well, if I'm pessimistic, I'll say a factor of 2 and, if I'm confident, 25%—without knowing *anything* about it except that it is not crystalline and not metallic.

HB: Well, I know nothing about this. Why is that? That seems quite strange.

TL: Exactly. *Why* is that?

Now there is a theory that's been around for roughly 40 years now, which basically says that in these materials, because they are disordered, there will always be places where a particular atom has available to it two possible sites, and then it can move between them by, say, quantum tunnelling and so forth.

That model is very generic and has lots and lots of free parameters. If you work hard enough, for any given material, you can find parameters that will give you the experimental values. But why should they be in that range?

That seems to be a case where there has only been this one model sitting there for the last 40 years and nothing else has been developed to the same degree of quantitative refinement. Therefore, experimentalists simply automatically fit their data to it.

HB: Is there only one model because not many other people are concerned about this? Or is there only one model because it's an effective model that seems to be working?

TL: I think both. Certainly the whole field of glass is not a particularly glamorous one and is not a major topic of interest in most parts of the world.

It's probably more so in Europe than in North America, but still, even there, it's not that popular. But also, I think most people feel that we have this standard model and it seems, right now, that there is no smoking gun evidence against it, so why not use it?

HB: Are there other materials that exhibit similar properties?

TL: This seems to be pretty much universal among the classes of really amorphous, that is, really glassy materials. Some disordered crystals seem to show the same kind of behaviour, and metallic glasses show some of these properties, but not all of them, as you'd expect, since they have an extra ingredient, as it were; that is, they have free electrons running around.

Questions for Discussion:

1. What does it mean, exactly, to have "lots and lots of free parameters"?

2. Are you surprised to learn that some scientific subjects are more or less popular in one geographical area than another? What do you think might be the factors responsible for that?

*3. How do you think Tony would respond to the following reply, "**Our only concern is to have a model that works**"?*

V. The Very Complex

Condensed matter physics meets quantum information

HB: Since we've moved to condensed matter physics, let's speak more generally about what's happened there since *The Problems of Physics* was first published. But let me first ask you to back up even further and discuss what you mean by 'condensed matter physics' in the first place.

TL: Well, I think there's a narrow definition of condensed matter physics and then there's a much broader one.

The narrow definition would essentially correspond to what in the old days used to be called "solid state physics"—plus, perhaps, a bit of "liquid state physics"—that is, matter under conditions of high density, reasonable temperature and so forth.

The broader definition would basically be that condensed matter physics is about any situation where you have a large number of entities that are strongly interacting in an interesting way.

For example, with that definition it would certainly cover much of ultra cold atoms. It would, by some definitions at least, cover cosmology because, after all, you've got a lot of matter interacting there. And in some people's definition it would also cover the stock market and things like that.

Quite a few condensed matter physicists have tried to take some of the techniques that were developed in their original work in condensed matter physics and apply them to things like the stock market, estimating the probability of its collapse and so on.

HB: Yes, well, I'm not sure if any have met with a tremendous amount of success there.

TL: That's probably a better opinion, I think, yes. After all, there are those who would attribute the collapse of 2008 to the mere influx of condensed matter physicists to the financial sector.

HB: Well, not just condensed matter physicists, we have to be fair: lots of other types of mathematical scientists were involved in all of that. And presumably regulation played a role there as well.

TL: Yes, or the lack thereof.

HB: Exactly.

So what's changed in condensed matter physics? What are the really exciting topics for you? This is very much in your ballpark, in terms of your areas of expertise. I'd still like you to speculate but perhaps you can speculate with even a little more authority now.

TL: To my mind, the most exciting thing that's happened in the general area of condensed matter physics recently has been the connection that has developed between that field and quantum information.

When the field of quantum information got off the ground, which I suppose one could place in the early 1990s, I think most people assumed—including me at the time—that the kind of systems which were going to be really interesting from the point of view of quantum information, quantum computing and so forth, would be simple microscopic systems: photons, trapped ions, possibly well-isolated nuclei, and so on.

And I think the *really* interesting development has been the proposal for topological quantum computing: that you could get around many of the problems which afflict quantum computing done with things like photons or trapped ions by using the complicated, entangled properties of many-body systems.

HB: Let me back up a little bit more. If I'm somebody who doesn't know anything about this, and I hear the words "quantum information", what should I be thinking about and why should I particularly care about it?

TL: Quantum information is a fairly broad field, which, in some sense, encompasses several different attempts to apply the bizarre features of quantum mechanics for engineering purposes.

Of course, for many decades now, there has been application at a certain level, like a transistor. But one's really talking now about things that, in some sense, are much more essential features of quantum mechanics: the basic phenomena of interference, entanglement, and so on—particularly entanglement.

According to quantum mechanics, it appears impossible, in certain circumstances, to give an adequate individual description of two objects that may have interacted in the past but are now physically separated.

That was the phenomenon that Erwin Schrödinger realized way back in 1935, but didn't know what to do with it, exactly.

John Bell, in some sense, put his finger on how one might verify— as far as one can *ever* verify anything in physics—that nature *does* behave in a way described by quantum mechanics in the kind of situation where entanglement is important.

It took quite some time but starting in perhaps the mid 80s and then really exploding in the mid 90s, people realized that there was a possibility of exploiting these bizarre features of quantum mechanics for real engineering purposes.

One purpose for which it's already exploited in real life is quantum cryptography. This is a subject which prior to—I think the first paper was in '84 but it really didn't come into its own until the mid 90s—simply wasn't a subject before. Now it's a huge subject and not only academics but commercial enterprises are interested in it.

HB: Why did it take so long for quantum cryptography and quantum computing to develop? Was this because the technology wasn't there or was this because, in your view, there was some reluctance to develop these ideas in some particular way?

TL: I think for a long period, in the early to middle 20th century, discussing foundational issues was sort of taboo. I think I'm correct in saying that between approximately 1935 and 1960 the number of conferences on the foundations of quantum mechanics that were held in any Anglo-Saxon country could probably be counted on the fingers of two hands. Somehow it was not a respectable thing to be doing.

It's interesting that the people who really, in my opinion, made a decisive change in that situation were either not fully professional physicists—people like Abner Shimony, who had a joint appointment in philosophy as well as physics—or they were physicists like John Bell, who were doing this in their spare time. John Bell did a lot of things in his spare time, including some very important work in theoretical particle physics.

So it really was these mavericks, as it were, who made the real impact in that field. Of course, eventually people realized the importance of Bell's work and by the late 70s quite a lot of people were starting to get interested in it.

HB: There seemed to have been this sociological taboo, then: this idea that Niels Bohr had somehow figured it all out without bothering to try to understand what Bohr actually said—which is a nightmarish task in and of itself.

TL: Indeed.

HB: And these issues were somehow ignored until...when, exactly? The mid-to-late 80s? The early 90s?

TL: It was actually later than that when things became respectable. I was interested in the foundations of quantum mechanics for quite some time. When I moved permanently to the US in 1983 I continued with that work and published a number of papers.

Sometime around the year 2000 or shortly thereafter, a group of us at Illinois wanted to make an application to the National Science Foundation for work that had some bearing on the foundation of quantum mechanics.

So we had to submit the usual documentation. And one of the pieces of information we were asked to supply was to cite any papers in the general area of the proposal that had been published with NSF support.

During my time at Illinois between 1983 and 2000 or so, I had probably published around 20 papers in that area. However, I had not even bothered to *try* to get NSF support for them because it was so unfashionable. So the pendulum has swung.

HB: You began this part of the conversation by saying that the key development in condensed matter physics, in your view, has been its evolving relationship with quantum information theory. Have there been other important and influential developments in condensed matter physics in the meantime?

TL: Nothing springs immediately to mind.

One thing that was actually discovered shortly before 1986, but has been naturally reverberating throughout the years, is the fractional quantum Hall effect. One could say that it's the most sophisticated problem in condensed matter physics that we can claim to have some sort of reasonable quantitative understanding of.

Whether we have a complete understanding or not is still not entirely clear but at least we can do a certain amount with it; we can make certain theoretical predictions which experiment does seem to bear out.

Additional Information:

Readers frustrated at the highly cursory nature of the discussion of concepts in this chapter are advised that a somewhat more detailed treatment of foundations of quantum theory will occur in Chapter 8.

VI. Understanding

What it actually means

HB: Let me pick up on that and ask you a little bit more about what it means to really understand something, what an explanation is in physics. Maybe this sounds like an obvious thing: you know something when you have a theory that can be repeatedly tested and consistently works. But to my mind there are many aspects of this that are actually quite a bit more complex, and you go into some of them in your book.

TL: Yes. I think in some sense we understand, or at least convince ourselves that we understand, a phenomenon in physics—a particular set of experiments in physics, say—if we can fit them into a general Gestalt theory. Of course the Gestalt can be extremely broad, it can be something like Newton's laws. Or it can be something much more specific: it can be a specific theory of how particular solids are constructed and what their electron energy bands are like and so forth. But we need to be able to relate that phenomenon to other things that we have observed in physics.

Of course one could probe deeper than that and ask, "*OK, let's take something like Newton's laws themselves; what does it mean to say we understand those? Does it even make sense to ask whether we understand them or not, or do we have to simply accept them as a fact of life?*"

This is a very interesting question in connection with the foundations of quantum mechanics. We have the formalism of quantum mechanics, which has now been around for almost a hundred years. We can apply it and—at least in most of the areas in which we try to apply it—it seems to work pretty well. There's no evidence, right now, that it fails at any particular point. But *why?* **Why** does nature work that way?

And one very interesting subfield of the foundations of quantum mechanics has been the attempt to derive quantum mechanics from more elementary, basic postulates.

The kind of postulates that people use are, for example, the inability to transmit information faster than the speed of light—Einstein's basic principle of special relativity—the continuity of the description of the state and so on.

I think that's a very interesting enterprise although, frankly, even if it succeeds, it won't necessarily make me anymore confident that quantum mechanics is the whole truth about the world.

HB: And even if it were to succeed, you might well have the same problems with those principles and those postulates.

TL: Yes. My experience with small children is that they're always asking, "*Why?*" You tell them, and they ask "*Why?*" again. Eventually you just have to throw up your hands and say, "*Well, it's just like that.*"

HB: As you were talking I thought of something else you had written in *The Problems of Physics*, which struck me as a particularly noteworthy insight.

Earlier in the book you speculate that if people in the 19th century—or perhaps even in the 18th century—had the computational power that we have now, they would have been cranking away at all of Newtonian mechanics and so forth and they wouldn't have developed some of the mathematical analytic techniques, in particular the principle of least action, which paved the way for the eventual reconceptualizing that people had to do once quantum mechanics came along.

In order words, the transition from classical mechanics to quantum mechanics would have been harder had they had access to effective computational machines.

That was a really intriguing idea to me: that, to some extent anyway, sitting back and calculating away impedes one's ability to develop the appropriate mental constructs that might be necessary to develop a better and deeper understanding of nature.

TL: Yes. That's why I'm not myself a tremendous fan of computational physics. Obviously there are plenty of places where it's not only useful but probably essential. But I slightly worry about it taking over more and more of physics.

Questions for Discussion:

1. To what extent is it reasonable to imagine that it's possible even in principle for physicists to come up with a "Theory of Everything"?

2. Might computational physics enable us to make fundamental advances in physics that we otherwise wouldn't?

VII. Different Regimes

Nature's scales

HB: Let's talk a little bit about what it means to be "fundamental". This is a word that gets thrown around an awful lot: "***This** is fundamental. **That** isn't truly foundational.*"

There's a long history of people making grandiose—there's that word again—statements about what it is to be foundational. And sometimes condensed matter physics winds up on the short end of the stick. Wasn't it Murray Gell-Mann who once called it "squalid state physics"?

TL: I think that's right, yes.

HB: Give me a sense of what your views are about all of this.

TL: Perhaps I can best define what I regard as "fundamental" by what it excludes. I think what it excludes is the kind of operation—which is very common, and all of us, including me, do it all the time—where one starts off from a set of assumptions which, at least for the purposes of the calculation, one is not going to challenge. One then tries to infer certain consequences of these basic assumptions.

A typical example might be what a lot of electronic band structure work is about. You start off with, basically, Schrödinger's equation for a set of 10^{23} atoms or whatever, and you try to make approximations to them.

Now, it's not obvious that some of this is not what I would call fundamental or foundational, because in the process of making those approximations you may have to, at some stage, introduce new concepts. To the extent that you're introducing new concepts, which can't be totally explained away in terms of the existing ones, then I think you are, to an extent anyway, doing fundamental work.

I think a very good example of what I would call fundamental or foundational work in condensed matter physics is the work of Lev Landau—in particular his beautiful work on superfluid helium, where he introduces the idea of elementary excitations, the idea of a normal component, a superfluid component and various other things. These were not things which you could define rigorously in terms of a simple picture of 10^{23} atoms or whatever.

HB: Often, I think, physicists associate "foundational", either deliberately or subconsciously, with a reductionistic approach: this notion that you take something apart and look at its constituent components to get to the core of it. But one doesn't have to be a physicist or a scientist to know that that's not the answer to all of our questions about the world.

We know that if we take ourselves apart we eventually get to these subatomic particles, and we know that if we take this table apart we also get to subatomic particles. But, of course, there's a fundamental difference between ourselves and this table—namely that we're alive and the table isn't—and that's just not encapsulated at all in this picture.

So even at a very coarse grain level, an intuitive level, we all realize that reductionism has its difficulties. But of course, you don't have to go that far: there are all sorts of levels of emergent structure that happen.

How much do you think being fundamental is linked to the spirit of reductionism? Should we be doing anything different about that? Should we be telling people from an earlier age so they have greater appreciation?

TL: As a professional condensed matter theorist, yes, I think we should.

That is, I think that the instinctive view, as you say, that you understand how things work by taking them apart and so forth, is an implicitly anthropocentric kind of view.

You have a radio or a bicycle or whatever, and if you're curious you might unscrew the back and take the various bits out to try to figure out how it works. But that whole approach might be misleading, because that radio has been put together by human beings: it's not so obvious that this is generally going to work on the objects that nature creates.

But even if you accept that, I frankly just *do not see* why the study of the microscopic components of a macroscopic object is more fundamental than the study of how these interact in subtle ways when they're all put together. This is the kind of point that Ilya Prigogine, Phil Anderson, Bob Laughlin and so forth have tried to make.

Personally I'd go even further, and this is a very radical and minority point of view, I think. I would claim that there is a real possibility that there may actually be laws of physics that *only* come in at the level of subtle, complicated, macroscopic objects.

In other words, I think my difference with people like Prigogine, Anderson, Laughlin and so forth, would go something like this: we all agree that even quite inanimate phenomena of various kinds *can't* be, in practice, reasonably explained—in any meaningful sense of the word "explained"—by simply writing down Newton's laws or Schrödinger's equation for 10^{23} particles.

But people like Laughlin and company would say that at least the phenomenon is *consistent* with the speculation or conjecture that Schrödinger's equation does work. It may not be a very useful piece of knowledge, but at least it's consistent.

I would go even further than that and say that I would not be totally surprised if there were actually *new laws of physics* that would come in at some level of macroscopicity or complexity which mean, crudely speaking, that quantum mechanics is not the whole truth about the world.

HB: So let me be devil's advocate and explore things a little bit here from the position of the hardcore reductionist.

I might say: "*Fine. We have some confidence in the laws over here; but you, Tony, say that at some level of macroscopicity new laws kick in.*

Well then, it seems like I would have to have some sort of mechanism for that, some sort of structure, maybe some sort of meta-structure that determines when I hit this level." How would you respond to that?

TL: I think I would respond to it with a sort of analogy.

Imagine that I'm a time traveller travelling back from today to the year 1870 or thereabouts. I meet a group of physicists and I assure them that, at some point, as you go down in scale from the order of the macroscopic world to the atomic world, the laws of mechanics are going to change radically and fundamentally.

They would look at me, somewhat perplexed, and say: *"OK, but **what's** going to change? After all, mechanics has no scale in it."*

And I would say: *"Well, sorry mate. Actually, according to modern information there **is** a scale."* It would not have been reasonable for them to guess it at the time, but it is there.

I would speculate—and in some sense hope—that even after the next major revolution, if there is one in physics, we'll still be able to use quantum mechanics for the purposes of describing phenomena at the atomic level, just as we now still use Newtonian mechanics to describe planetary motion. Nothing has gone wrong with that.

HB: But at some point we'll be in a different regime.

TL: Yes. And my reasons for this are not totally arbitrary. Again, I'll make a historical analogy.

Project yourself back, not to 1870 but to 1875. What's special about that year? In 1875, at least as far as I know, there was no real evidence in atomic physics or optics of a failure of classical mechanics or classical electromagnetic theory at the experimental level. However, in that year, Gibbs published a paper that included the so-called "Gibbs paradox".

Crudely speaking, you can think of mixing two bodies of gas, and if these are different gases then all the consequences that arise concerning the entropy of the final, mixed state of them seem sensible.

If you do it for two bodies of identical gas, then—and here's the crux—if you assume that you could put separate markers on the distinct bodies of gas, then you get into various technical difficulties concerning the so-called "extensivity", the dependence of the entropy on volume, and so forth.

To people in Gibbs' day, the idea that you couldn't just put a mark on a particular atom would have seemed very weird. I think one can claim, with the virtue of hindsight, that if Gibbs and his contemporaries had taken that paradox seriously, they would have reached the conclusion that at some scale, as you go down from the level of the macroscopic to the atomic, classical mechanics is going to break down in some way.

They would not have been able to infer *at what point* it would break down, and even less would they have been able to infer *in what way* it would break down. But *that* it would break down, is something that I think, in retrospect, they could have inferred.

And I think we're in the same position with quantum mechanics today. I take the quantum measurement paradox, or the Schrödinger's cat paradox, sufficiently seriously to believe that. Again, we can't reasonably infer *the point* at which quantum mechanics is going to break down. Still less can we infer *the way* in which it's going to break down but *that* it's going to break down, I think we can infer that.

HB: So I'd like to get to that, right after I ask this question.

Earlier, you highlighted the growing overlap between quantum information theory and condensed matter physics as a very positive advance—perhaps the key recent advance for condensed matter physics in the past generation.

Meanwhile, there are some who believe that the foundations of quantum theory are extremely important, that these questions have been swept under the rug for a very long time and need to be addressed, but are deeply sceptical that quantum information theory—often encapsulated simply in "quantum computing"—is going to do anything to advance that project.

Their argument is typically along the lines of, "*This is just engineering. It's like developing transistors and all the rest of that. That's all fine and good, and important in terms of GDP and so forth, but I'm a foundational physicist, so don't tell me that this is going to address any core issues in the foundations of quantum theory, because it's not."* How would you respond to that?

TL: Well, I think that depends on whether attempts to devise realistic versions of quantum computing work or not. If indeed, as people go to more and more entangled states, to more and more macroscopic qubits—flux qubits and so forth—if everything still works fine according to the basic prescriptions of quantum mechanics, then I think we're not going to learn anything.

The more exciting speculation is that, as we go in this direction—it's just one of the many directions we might try to go, but it's one frontier along which we can explore the continued validity of quantum mechanics—it just could be, because of the particular entanglement properties of some of these very sophisticated states we'll need, that may be the trigger for quantum mechanics to break down. There's a very small probability of that right now, but it could happen. And in that case, we'd have learned a lot.

HB: So it's not so much that we may stumble upon something as a result of the theoretical constructs we're developing within a quantum computational framework, but rather by playing, by pushing.

TL: We may, but I stress the "may".

A phrase which I've encountered many, many times when I review proposals and so forth is: *"By doing this or that experiment we'll be exploring the boundary between the quantum and the classical."* Nonsense. If it works according to the prescription of quantum mechanics, you've learned *nothing* about this. If it breaks down however, there's a *huge* amount that can be learned.

Questions for Discussion:

*1. To what extent do you think that Tony's analogy of a "time-travelling physicist" represents a "thought experiment"? Interested readers might wish to compare it to Nima Arkani-Hamed's time-travelling analogy in Chapter 14 of the Ideas Roadshow conversation **The Power of Principles: Physics Revealed**.*

2. In what ways does Tony's example of the Gibbs Paradox provide a form of evidence for the limitation of our theoretical framework independent of any experiment?

VIII. Schrödinger's Cat

Different domains?

HB: I can imagine people listening to you who don't have a technical background thinking: "*Here's Professor Leggett, an eminent physicist who has clearly had a distinguished career—he's even won a Nobel Prize. He seems to be upset about this thing called "the foundations of quantum theory'", yet at the same time I'm hearing that the theory works: it's predicting what it should be predicting."*

What's bothering you, exactly? Let's talk about what you mean by "the foundations of quantum theory" and try to get at why you're so restless about this.

TL: What really worries me is Schrödinger's cat. I'm less concerned about the whole EPR-Bell setup, except insofar as that's related to Schrödinger's cat, and of course it *is* related at a deep level.

HB: Tell us, first of all, what Schrödinger's cat is.

TL: In Schrödinger's original example, you have some sort of microscopic system—in his case it was a radioactive device but we might, for example, think of a photon which can have, classically speaking, vertical polarization or horizontal polarization.

If it has a vertical polarization then it will pass an appropriate, properly set polarizer, and if it has a horizontal polarization, it will be rejected.

Suppose that the photon is rejected, then nothing much is going to change in the world. But suppose it's transmitted. Then we set up a complicated device—a "hellish device" as Schrödinger called it—behind the polarizer, which will trigger various electronics, and as a result of this, a cat, which is sitting innocently inside an enclosed box, will be killed.

If the first photon that we fire in definitely has horizontal polarization, then the cat will live. If the photon definitely has vertical polarization then the cat will die. However, unfortunately quantum mechanics allows the possibility that we have a so-called quantum superposition of vertical and horizontal polarization, and we can show by various subsidiary experiments that this is *not* equivalent to a straightforward mixture of these two: that is, it's **not** the case that the photon was definitely vertical with some probability or definitely horizontal with a different probability.

Suppose we send this 45-degree, polarized photon into the apparatus. What does quantum mechanics tell us is then going to be the future state of the universe? Well, quantum mechanics is, by its construction, a strictly linear theory. So if the initial state was a quantum superposition of the two different polarities, then the final state of the universe is going to be a quantum superposition of a state in which the cat is dead and one in which it's alive.

This is a description that we naturally find very difficult to interpret.

At the microscopic level, there are all sorts of subsidiary experiments we can do to convince ourselves that the superposition is not equivalent to either one state or the other—it's more sophisticated than that.

At the macroscopic level on the other hand, we're used to taking the lid off the box to see if the cat is either dead or alive.

Perhaps the situation is a little more intuitive if we take the famous two-slit experiment. You have an electron that has available to it two paths going through two different slits, and then you do an interference experiment.

I would prefer, in that situation, not to make any positive statements like 'it did go through both' or 'it didn't go through both'.

But I can make the negative statement which is **not** correct, that each individual electron *either* went through slit A *or* went through slit B. That's one thing I **can't** say.

This is the point which, it seems to me, much of the literature on decoherence seems to miss. Decoherence is a technical trick for pretending to have solved the measurement problem. The formulas of quantum mechanics haven't changed a whit as we go from the description it gives of the photon to the description it gives of the cat. If we refuse to make a particular interpretation at the microscopic level then we have no business reintroducing that interpretation at the macroscopic level.

The evidence for this statement is there at the microscopic level, in the form of interference patterns, but everyone agrees that, at the level of the cat, it has gone away.

But does the fact that the evidence against a particular interpretation of the formalism has gone away by the time we get to the cat mean that we can freely reintroduce that interpretation? I say, no.

HB: What works for the photon has to work for the cat. And vice-versa presumably.

TL: Right.

HB: Of course that was Schrödinger's whole point, presumably: to ramp things up to a level of absurdity. We're happy to talk about photons in this particular way, but we consider it absurd when talking about cats.

TL: Yes.

HB: And this gets back to what you were saying earlier: if new rules come into play, where does that happen? How does that happen?

TL: Exactly, yes. When I first started thinking seriously about this, way back around 1980, I really quite seriously hoped that when you got to the level of the so called "flux qubit"—where the two states you're dealing with are different in the behaviour of something like let's say 10 billion electrons—I was rather hoping that by that time something else might have happened. Right now it looks as if quantum mechanics is working fine at that level.

HB: But isn't there another issue in terms of internal consistency? This is something that has always bothered me. I'm referring not only to the language that is used to describe this, but also the theoretical framework itself.

The theoretical framework incorporates this notion of a measurement. And this notion of a measurement, it seems to me, is not a terribly well-defined one that necessarily extends outside of the system. For a lot of people that's yet another problem with quantum cosmology: who's going to be measuring this wave function of the entire universe? And what does that even really mean?

But just to return to ordinary quantum theory, you have this state vector or this wave function which is some description of the system, and then there's a question about how you use that, how you interpret that, how you extract information from that.

But it seems to me that there is an additional problem: not only are we uncertain what that actually means, but to extract information structurally we have to have something that is outside of the formation of our theory.

Do you look at it that way? Does that bother you too? Or is that somehow part and parcel of the whole issue?

TL: I think it's really part of the original paradox, yes. The fact that we are, as human beings, conscious of seeing definite outcomes—in fact we can't even imagine what it would be like *not* to see an outcome— whereas the formulas of quantum mechanics in some sense tell us that we shouldn't be.

Questions for Discussion:

1. To what extent do we have to fully understand our theories of nature in order to embrace them?

2. Do you think that the Schrödinger's cat thought experiment demonstrates the limits of a reductionistic world-view"?

IX. The Slings and Arrows of Time

Irreversible?

HB: When we last met, you mentioned that one of the things that you had been musing about was how issues related to the arrow of time might somehow be associated with some of these foundational issues of quantum theory.

I'd like you to indulge me yet again and speculate a bit more about that.

TL: OK. Of course the issue of the arrow of time has been around for a long time and a lot of people have thought about it. One thing I'm relatively happy with—and it's part of the common wisdom on the subject—is that if you're going to look for an ultimate explanation of the arrow of time, in the end it has to go back to cosmology.

The most general aspect of the arrow of time which we're conscious of in everyday life can be summed up in the Second Law of Thermodynamics: that disorder tends to increase as a function of time.

Given that general principle, I think it's not so difficult to extract particular applications; for example, that one can remember the past and affect the future and so forth. I'm not saying I can do it—and I think that people have perhaps tended to go overboard in saying that they know how to do it—but I can't see any obvious insuperable objections to doing it. It seems reasonable.

However, how are we going to justify the thermodynamic arrow of time? One possibility is to say: "*Well, that's just the way it is. Entropy was low in the past and is increasing as we go into the future. That's just a fact of life.*"

If you don't like that then I think the obvious solution is to look at cosmology and acknowledge that entropy has to do with the fact that the universe is expanding from the hot Big Bang where the entropy was very low. Why? Well, that's a huge question and all the heavy-weights of the field have tangled with that and fought over it. I'm not necessarily going to go into that (*the reader who **does** want to get into it, however, is referred to the Ideas Roadshow conversation **The Cyclic Universe** with Roger Penrose, where the question is discussed at length*).

But let's just suppose that problem is solved, I think you *still* have a problem with the following question: given that the overall arrow of time is in the direction that we all know and love, is it possible that there are fluctuations—and here I'm really talking off the top of my head—in small regions of space-time in which, in some sense, the normal arrow of time would be reversed—in particular, in which one could legitimately ascribe the cause of an event to what is going to happen in the future, instead of what has happened in the past?

HB: These are small fluctuations, presumably?

TL: Well, for the moment at least, let's say so.

One possible reason for taking that speculation slightly more seriously than one might otherwise do is the EPR Bell experiments. You can look at the raw experimental outputs of these experiments without ever mentioning quantum mechanics. Just look at what the outcomes of these experiments tell you. They tell you that you've got to reject at least one of three things which, at least in everyday life, we know and love.

One is our normal, core concepts about the arrow of time: that, in some sense, we can remember the past and affect the future; that events in the future cannot causally affect events in the present, etc.

The second is locality in the technical sense of special relativity—OK, not everyday life but physicists really love that principle.

And the third, well, you can put it in various forms, but the way I like to put it is "macroscopic counterfactual definiteness"—the idea that you can assign truth-values to statements, conditional on unperformed conditions.

We do this all the time in everyday life. I say to you: *"Had I got up 5 minutes earlier this morning I would have caught the bus."* You don't question that that's a meaningful statement.

Not only do we make these counterfactual statements and think of them as meaningful in common, everyday life, but the legal system absolutely hinges on these counterfactual conditions. The prosecuting counsel says to the jury: *"Had the accused not pushed the victim down the stairs, she would still be alive today."* The jury has to make a decision whether they believe that statement is right or wrong. So they certainly take it as a factual statement.

HB: Right. So there are these three fundamental aspects.

TL: Yes: these three common sense assumptions, one of which has to go. I'm not happy about throwing away any of these, but I think we should at least think about the possibility of throwing away our normal assumptions about the arrow of time.

This would allow us to regard the actual outcome, the measurement at one end of the lab, as in some sense determining the initial state of the pairs of photons that were emitted at the beginning, even though the measurement of course occurred later in time.

HB: Perhaps I'm just being closed-minded or stubborn but I'm trying to get my head around this. The second one is special relativity: information can't be transmitted faster than light, that's no problem. But let's suppose we throw away the first one, the causal one, which states that things from the past affect the future and you can't go the other way, right?

But I'm keeping the third one, which is this counterfactual claim that had I not pushed this person down the stairs then they would still be alive. But it seems to me, implicit in that had I not pushed them down the stairs is causally linked to them still being alive.

TL: Well yes, I think that's right. But you are linking it to our normal ideas about causality: the push came before the death, or whatever.

HB: I see. So the idea is that dying could have "caused" being pushed down the stairs.

TL: That would be a denial of the first assumption, yes.

HB: So we still have special relativity—which thankfully doesn't have to do with stairs or any of that—and I have a logically consistent framework whereby I die which causes me to then be pushed down the stairs.

TL: Well, yes. But that's separate from the case of what would have happened had he not pushed me down the stairs, whether it's even meaningful to attach a truth-value to that statement.

HB: OK, I see: whether it's even meaningful to attach a truth-value to that—your "macroscopic counterfactual definiteness".

OK. So that's the logical framework. Now how does abandoning the first of these three—namely that things from the past causally affect things in the future—how does that somehow help or rescue or shed light on things?

TL: Well, very crudely. Let's go back to the original formulation of John Bell's argument, which has been, of course, generalized a lot since then. Basically the idea is that you have some distribution of parameters that describe the pairs of photons that are emitted from the source and, crucially, this distribution cannot depend on the setting of the instruments that are being used to measure them.

The reason that one would normally give for that has to do with the fact that you could choose the setting at the very last moment, as Aspect and others actually did in their experiments.

Therefore, it's in a future light cone, technically: the future of the event of emission and therefore can't affect it. If you were to reject the normal assumption about the arrow of time then that conventional understanding of causality would disappear and you could say that the setting of the instruments and/or the outcome of the measurements could actually come in from the future and affect that distribution.

HB: I understand how changing the arrow of time will affect my sense of causality. I understand how affecting my sense of causality will shine new light on the EPR experiments by allowing for the measurements of polarization to somehow "cause" the state of the photons rather than the other way around, which is how we normally think of it.

But it seems to me that if what you're saying is true, you need to be saying something more sweeping about the arrow of time. If this is happening all the time you should be able to go into a lab and do experiments over and over again that show this. I mean, I'm guessing that this phenomenon isn't happening once or twice in a millennium: it's happening all the time.

If the arrow of time can go backwards, if you change the arrow of time, I understand how that changes everything in terms of my interpretation. I'm willing to believe—I think—that every so often, somehow, in some small context, the arrow of time can go backwards, but when I do these experiments, everything else in my life seems to be going forwards.

TL: Maybe that's because you're a macroscopic being. This is completely off the top of my head, but think about the timescale that's involved here.

In the most recent Guinness Book of Records, I think the EPR Bell experiments had a spatial interval of something like 100 kilometres. So what does that work out to in terms of time for light transmission? Something like a few microseconds, perhaps a little more than that. But anyway, that timescale is incredibly short compared to the time-scale of human consciousness. So, who knows? Speculating wildly, maybe these little reversals at that kind of level are going on around us all the time and we're just not aware of it.

HB: Again it's definitely worth emphasizing that I'm the one who is forcing you not only to speculate, but to do so publicly and off the top of your head to boot.

But would you imagine that these reversals are consistent at that particular time scale to the extent that we could keep repeating these experiments over and over again and we would always find that?

Or would it be likely, but not necessarily the case? So we would keep doing those experiments because every so often we're going to find that these reversals are not actually happening?

TL: I don't know, quite frankly. That's a very interesting question, and it is perhaps somewhat related to a much more general question which I think is very frustrating and very slippery: *Is there any kind of framework such that you could embed this notion of temporary and local reversal of the arrow of time in a larger framework so that the Second Law of Thermodynamics would continue to work?*

I just don't know. I always say that the most difficult questions in physics—I suppose this is a more general statement, actually— are not the ones where the questions are well posed and you don't know the answer. The *really* difficult issues are the ones where you don't even know what *questions* you should be asking. I think this is precisely one of these cases: it's just so difficult to formulate a meaningful question in this area.

HB: Do you feel that you're getting closer, at some level? Do you get a sense that you're converging on things in some particular respect?

TL: Frankly, no. I think my bet would be that sometime in the next 100 or 200 years or so there's going to be one more major revolution—at least one more—in physics; and my bet would be that it is somehow going to involve the arrow of time.

If you think back on the really big revolutions in physics like Copernicus, Einstein, Heisenberg, etc., every one of them has involved the abandonment of some principle about the world, which up to that time had seemed like the most basic common sense.

And the one bit of common sense that has *not* been seriously challenged is the idea that the past causes the present and the present causes the future. I'm just wondering if that's not the next one that's going to have to go.

HB: Well, it's the only one left.

TL: In some sense, yes, I think that that's right. Everything else that we thought was common sense has gone out the window.

HB: When you were talking about dark energy, you were speculating that it was somehow related to this pre-revolutionary Kuhnian state. Was it also your instinct that dark energy might somehow be linked to this notion of the arrow of time?

TL: I think it could go either way. Right now there do seem to be serious problems in the area of cosmology, which people are energetically putting Band-Aids on. This has really been the situation for the past 15 years or so in cosmology. There has also been this ongoing irritant, the Schrödinger's cat paradox, which is now almost 80 years old. Whether the two have any relationship, I don't know. But my suspicion is probably not.

Questions for Discussion:

1. Might an underlying theory of nature be beyond the human capacity for understanding it?

2. To what extent do you think it could be successfully argued that the very nature of a "counterfactual" claim implicitly relies upon the arrow of time?

X. The Anthropic Principle

Better left unsaid?

HB: I'd like to talk a little bit about the Anthropic Principle now because we haven't mentioned that yet. We seem to be hitting most of the big foundational issues, so we might as well hit that one as well.

TL: Sure. Why not?

HB: Once again *The Problems of Physics* seemed particularly prescient to me when I recently read it, although perhaps that's simply because I was fast asleep in 1987 and was a lot less sensitive to what was happening then than you were. At any rate, you go on at some length about the anthropic principle. It's one of the "three skeletons in the closet" that you outlined.

TL: Yes, I think the other two were the arrow of time and the measurement problem.

HB: Right. So we've almost covered them all, but now we've got to go back to the anthropic principle. A couple of things struck me as very interesting.

The first was that, back in 1987—in fact a lot later than that too—particle physicists were very ambitiously proclaiming that their theories would be able to put the anthropic principle to bed once and for all through the development of their so-called "theory of everything".

The idea was that, once everything would be worked out properly there would be no more free parameters and everything would be determined by this all-encompassing theoretical framework. People like Stephen Hawking, if memory served—although he was by no means the only one—famously proclaimed that the theory of everything was just around the corner.

Nowadays, however, many of the very same people who made such proclamations with such confidence have not only abandoned those claims, but they've abandoned them *within the framework of the anthropic principle itself.*

Back then they were saying that they were going to kill the anthropic principle by developing a theory of everything; now they're saying that they *can't* develop a theory of everything *because* of the anthropic principle. Does that strike you as ironic?

TL: Yes, it does.

HB: And what do you think, personally, of the anthropic principle?

TL: The reason that people nowadays, as distinct from the 17th century, are interested in what we call the anthropic principle, is that in the current model of particle physics there are many numbers that don't seem to have any prior explanation.

A typical example would be the ratio of the mass of the electron to the mass of the proton, or the electron charge in appropriate dimensionless units. Right now, the so-called standard model of particle physics has to just put those values in by hand and say: *"That's the way it is, and we don't have any explanation of that."*

For a long time it's been observed that, not only the existence of human life and therefore presumably human consciousness, but even much coarser features of the universe—like the existence of galaxies—are extremely sensitive to the values of these parameters.

People claim—I can't judge the accuracy of this claim—that if you were to change the dimensionless charge of the electrons by 1 part in 10^8 or 10^9, what we know as chemistry would be quite different and it's improbable that human life could have evolved under those conditions.

We can pinpoint two extreme versions of the anthropic principle in this context.

One extreme version would be to say that God or some extra-terrestrial influence deliberately picked and chose these fundamental constants to have these values in order that human life should evolve.

The other extreme would be to say that the structure of the universe is such that, in fact, these constants could have any values whatsoever, and perhaps there are different spatial regions of the universe in which they *do* have all possible values.

But if they *had* values very different from what they are, we would not be here to ask the question. So in some sense it's not an accident that, say, the value of the electron charge is what it is— simply because, were it not so, we would not be here at all.

I think the first version would appeal to people with certain kinds of religious beliefs but would not appeal to most physicists *qua* physicists.

The second version I think is more conjectural. I don't think it's totally absurd, but it's one of those questions that seems impossible to get any evidence about, for or against, experimentally.

HB: Earlier in our conversation you referred to yourself as a "terrible Popperian" or words to that effect. And as you know, many physicists have been upset by the invocation of the anthropic principle as an explanation because they say something like: "*It's not science, it's just an excuse for **not** coming up with a scientific theory. You're just saying: 'It is the way it is because were it not that way we wouldn't be asking that question', and that's not an answer. That's not the business of doing science. The business of doing science is focused on coming up with a real framework, a real structure to explain things*"— which gets back to this whole idea we spoke about earlier of what it actually means to explain something.

In your view is there something to that? Is the anthropic principle an explanation? Or is it something to be avoided at all costs?

TL: The question of whether or not, at the philosophical level, it's an explanation is a difficult one. An easier question is whether I would advocate it at the pragmatic level. And I would **not**, for essentially the reason that you said: if you accept that kind of explanation, then your motivation for looking within the structure of the existing theory for deeper reasons why these mass ratios should be what they are and so forth is very much weakened. That's a bad thing, in that it reduces the probability of advances in particle physics.

So even if I believed the anthropic principle was true, I would keep quiet about it among my colleagues.

HB: It would lead to lazy physics.

TL: In some sense, yes.

Questions for Discussion:

1. What do you think Tony has in mind, exactly, when he refers to himself as a "Popperian" (terrible or otherwise)? Those interested in more background on Karl Popper and his famous falsifiability criterion are referred to Chapter 8 of the Ideas Roadshow conversation **Science and Pseudoscience** *with historian Michael Gordin.*

2. Is it possible, even in principle, to objectively demonstrate the validity of the Anthropic Principle in certain contexts?

XI. The Future of Physics

From Louis Armstrong to topological quantum computing

HB: You've indulged me by speculating a great deal and I like to thank you very much for that, because that's much more fun than to hear someone go on about matters that he or she can passionately backup and verify. I think people rarely get the opportunity to see someone of your stature freely indulge in such speculations, and I'd very much like you to continue.

Let's say we have this conversation again ten years from now—and I very much hope that we do, by the way. What do you think might be new and different then? What will we have discovered, or come to understand better, over the course of the next ten years?

TL: Oh boy. Well, you know the saying attributed to the jazz musician Louis Armstrong? Someone asked him where jazz was going and his answer was: *"Man, if I knew where jazz was going I'd be there already."*

That's sort of a generic answer. But are there questions that we ask today to which we will know the answer in ten years' time–

HB: No, that's not my question. I don't want you to play it safe. I want you to go way out on a ledge and indulge me with what you think might happen with no evidence whatsoever: full, frontal speculation.

TL: I think ten years is difficult; can I speculate on 50 years from now?

HB: 50? Yes, you can do 50. Although I'm not sure we'll be able to do this again in 50 years.

TL: In 50 years, I think there will have been a major revolution in cosmology. I think there's a small but non-zero chance that we will have pushed quantum mechanics in the direction of the macroscopic world to the point where it will fail and break down.

HB: Hold on. You're starting to hedge your bets again. Do you think that quantum mechanics will break down at some level?

TL: Yes, absolutely. But if you ask me the probability of it breaking down in the next 50 years, then I would say it would be pretty small.

HB: But quantum mechanics *will* break down at some point.

TL: Yes, I believe it will. One other prediction I think I can make with a certain degree of confidence is that we will have found—by going to lower and lower temperatures, with better and better noise control and so forth—many-body systems which are orders of magnitudes more sophisticated than what we know at present.

HB: What effect will that have? Technologically, what will that imply? Will we be able to build a quantum computer that...

TL: Whether or not we'll be able to build a quantum computer may hinge on that, but I think it's more likely to hinge on much more bread and butter issues: can we, for example, engineer trapped ions—one of the favourite candidates for a quantum computer—can we engineer them to the point where we can address them essentially perfectly but nature can't interfere with our operations at all?

I think the issues involved in building a quantum computer right now are really engineering issues.

HB: OK, but I diverted you. What are the implications of the many-body systems that you were talking about?

TL: I'm not sure but I think it may give us a new way of looking at many-particle entanglement.

Right now there's a fairly standard way of looking at that, which seems to be perfectly satisfactory for the kinds of purposes for which we need it today.

I think, more likely, it's going to create new phenomenological—I hate the word "emergent" but perhaps this is a case where one ought to use it—concepts comparable to the idea of quasi-particles with fractional charge statistic, which came out of the fractional quantum Hall effect. But, again, we can't foresee what kinds of concepts these will be. But that would be my guess.

HB: And what advice would you give to a keen foundational physicist today? Someone who is perhaps 24 or 25 years of age, perhaps she has just got her PhD and is motivated to explore foundational issues. What advice would you give?

TL: Well, this is my own personal prejudice coloured by my own background, but I would encourage her not to stick with the clean, well-defined, microscopic systems that have been the bread and butter of quantum information for the past 25 years, but rather to seriously explore these more messy, complicated, many-body systems in the hope of getting something useful out of them, as I think we have with topological quantum computing.

HB: You mentioned that earlier. Could you say a little bit more about that?

TL: Whether you would consider it fundamental or not, I don't know. But what it has been is an incredibly interesting way of uniting issues in traditional condensed matter physics with issues in quantum information.

The idea is, crudely speaking, that you avoid the decoherence—the fact that the environment is always trying to mess you up, which afflicts standard quantum computing—by embedding your information in non-local properties of many-body systems.

The standard sort of idea is that you might be able to think in terms of fractional excitations of your many-body system, which are separated by large distances, which you can then braid—wind around one another and so forth—in such a way as to affect transformations in your quantum system.

I think it's an incredibly fascinating idea and it has led to an improved understanding, not only of some aspects of quantum information, but of some aspects of more traditional many-bodied physics.

I think that's where at least some of the action is today, and will probably continue for the next two or three decades at least.

HB: Great. Anything I haven't asked? Anything you'd care to add?

TL: I can't really think of anything, no.

HB: Well, thanks a lot, Tony. It's been a lot of fun and you've been a wonderful sport.

TL: You're most welcome.

Questions for Discussion:

1. Why do you think that Tony hates the word "emergent"?

2. Do you agree with Tony that quantum mechanics will "break down" eventually? If so, what effect, if any, do you think that might have on future science and technology?

Continuing the Conversation

The reader is naturally encouraged to read Tony's book, *The Problems of Physics*, in order to get the fullest possible perspective on this conversation

The Power of Principles

Physics Revealed

A conversation with Nima Arkani-Hamed

Introduction

Beyond nymphs, dryads and leprechauns

How does science work?

Well, that's easy, right? We start off by collecting information about the world around us and then we try to make some sense of it: we look for patterns, search for a more general understanding.

In time, we develop sufficient awareness of these patterns that we begin making predictions about what else might be out there: explicitly formulating hypotheses of what we would expect to see under specific, controlled scenarios. Then we go ahead and rigorously test our hypotheses by explicitly creating these particular conditions; coolly assessing whether or not what has been predicted does, in fact, occur.

If it doesn't, or at least doesn't with any sense of regularity or precisely in the way that we had envisioned, we'll be forced to accept that at least one of our original hypotheses was incorrect and head back to the drawing board to modify things in an attempt to develop a more accurate level of understanding.

Meanwhile, if all of our predictions do come true, then we'll find ourselves with increasing confidence in our understanding. At some point, we'll likely start calling it something more grandiose, like a "theory"; and if it keeps working like clockwork for everything in its applicable domain that we encounter, we will eventually be tempted to call it a "law".

Such is, in a nutshell, what most of us mean by "the scientific method". It is famously objective, logical and eminently reliable; and the fruits of its success, both pure and applied, are the single most obvious

factor in distinguishing the varying levels of progress between different human societies throughout history.

But what happens when the experimental arena becomes less and
less accessible? In the world of high-energy physics, say, where hugely
expensive laboratory facilities take decades to construct, what can we
do in the meantime towards uncovering nature's secrets? And how
might we conceivably make progress and build upon our knowledge
when no further experiments are in sight?

Can we just sit back and invent any hypothesis we want, secure in
the knowledge that, far away from any experimental arbiter, there
is no way of distinguishing, even in principle, between different
theoretical possibilities? Does science at this point simply become
science fiction?

To Nima Arkani-Hamed, faculty member of the Institute for Advanced
Study and one of the world's foremost theoretical particle physicists, that sort of talk is not only wrong, it demonstrates a profound
misunderstanding of all that fundamental physics has accomplished
since Galileo brought us into the modern era more than four centuries ago.

Arkani-Hamed emphasizes that our understanding of the physical
world is based on two phenomenally successful general principles
of 20th-century physics—relativity and quantum mechanics—which,
when put together, "almost completely dictate what the world around
us can possibly look like".

> *"You could easily imagine a world without relativity and you could
> easily imagine a world without quantum mechanics. Either way
> things would be tremendously less constrained. It's the existence of
> both of them that makes things so complicated.*

> *"In fact, if I were God and I was given principles—or a sub-God, given
> the principles of relativity and quantum mechanics by the actual
> God who told me, 'Now go, build a world,' I'd say, 'Sorry, can't do
> it. This is just impossible.'*

"They seem almost completely incompatible with each other."

And it is precisely this near-incompatibility, this overpowering rigidity resulting from the necessity of finding common ground between these two general principles, that provides a huge constraint to any successful underlying theoretical framework that we might conceivably imagine.

Which means that even without one single experiment we have almost no wiggle room to come up with truly innovative approaches, a state of affairs that almost never gets communicated to the non-scientist.

"I think that that's the thing which isn't appreciated by the general public: the rigidity. Instead, there is some sense that theoretical physicists, unencumbered by data from experiment, are just out there inventing leprechauns and fairies and nymphs and dryads around every corner; and every crazy idea you have is something that can be put out there. And while we all agree that experiment ultimately decides, until it does, anything goes: leprechauns and nymphs and dryads are all on the same footing.

*"But it just is not like that. The incredible rigidity that we have in our framework makes it **almost impossible** to come up with a new idea. It's very hard to modify things in any way without ruining everything."*

Appreciating the importance of these constraints and why today's theorists have such overwhelming confidence in these two guiding principles not only explains how physicists operate but also gives deeper insight into misunderstandings at the core of various public controversies, like the erroneous case of the faster than light neutrinos that burst onto the scene in 2011.

One of the things that infuriated Arkani-Hamed the most about the whole incident was how it demonstrated a near-total lack of public understanding of how contemporary fundamental physics is done.

"How was it reported in the popular press? Here is this incredible thing, we were told—you can go faster than light—and those physicists who were sceptical of the results were largely described as

people refusing to challenge the orthodoxy of big old Al, looking down on us, wagging his finger and telling us, 'Don't you dare go faster than light!'"

But that, Arkani-Hamed tells us, is laughingly far from the truth. Indeed, precisely because of the preeminent role that relativity plays in our understanding of the world, many theoretical physicists had spent an enormous amount of time and trouble years beforehand explicitly investigating whether it would be somehow possible to violate its effects. And what they had found, after painstaking effort, was in direct contrast to what these experiments were implying.

"The reason why all of us were sure this result had to be wrong was not, therefore, because we had never entertained the possibility that Einstein might be wrong. Exactly the opposite! **That's** *what's so frustrating about it. We had entertained it so well, we had thought about it so much, so responsibly and in such detail, that we* **knew** *it was* **impossible** *to have an effect as humongous as they found. One part in a hundred thousand sounds like a small effect, but all our previous work on violations of special relativity showed results that were much more stringent: one part in ten to the ten, one part in ten to the fifteen, one part in ten to the twenty—just way off from these humongous-size effects that they were finding. So that was a big source of frustration. We knew it had to be wrong.*

"People assumed that we were convinced it was wrong because we didn't want to question these underlying principles. Well, that's true in the sense that we know that you give up a lot if you do it. But we're not ideologues: we prepare for the possibility and study matters in so much detail that we know it cannot possibly be right, compatible with all the other experimental results we've had all this time."

Understandably, Nima Arkani-Hamed is anxious to set the record straight about a specific scientific incident while detailing common frustrations that were typically overlooked by the trivializing mass media anxious for a juicy story. But there's another point, too, he's keen to make: that all too often the popular characterization of how the scientific process works is deeply distorted.

"I think that it's important for people to understand essential aspects of how this sort of science is actually done. No one is out to suppress rebellious ideas. Quite the contrary: if you can find some even moderately rebellious ideas that have even a modicum of truth associated with them, that's the way you make your name in the field.

"But again, it needs to be emphasized that we're in this very, very tight straitjacket. We're not going to, at the drop of a hat, destroy this entire incredible structure that we've built up over four centuries which has served us so well unless there's a really good reason for it.

"Almost always, experimental or theoretical challenges to the structure are bound to fail. You shouldn't be surprised that they fail. And people shouldn't take scepticism as evidence of turf protection. It's really evidence of the great fact that we have this entire castle that we built over centuries that works so well."

So what of the future? How can researchers like Nima Arkani-Hamed somehow wriggle out of the formidable restrictions imposed by our current frameworks and move on to unlock even deeper secrets of the universe? By trying to find a new light in which to examine these very same underlying principles:

"I actually think one of the big things that should happen in the 21st century, the next really big thing we have to understand in fundamental physics, is that we have to understand where space-time comes from in a more fundamental sense.

"And one of the things that we will hopefully understand, when we know more how this works, is why it is that these two big ideas of relativity and quantum mechanics seem to both fight each other so much, on the one hand, which is why the world is so constrained. But on the other hand, they also go so wonderfully together, in the sense that in a world that did not have quantum mechanics, it would be easy to imagine modifying relativity somehow. Similarly, in a world without relativity, it's really easy to modify quantum mechanics.

"Each one buttresses the other in a very strange yet really remarkable way. Why these two principles fit together in the exact way that they do is one of the greatest mysteries."

To today's leading theorists, the two founding principles of 20th century physics are simultaneously a constraint and an inspiration: a straitjacket and a guidepost for future discoveries.

The future awaits.

The Conversation

I. Physics Time Management

Giving it your all

HB: I've had this question that's been rattling around in my head for some time. I go into the bookstore and look at popular science books, and I ask myself, *Are there people who should be writing a popular science book who aren't doing so?*

It seems to me that there's a bit of disconnect between what the general public is aware of through popular science books and what's really going on at the front lines, who's really driving a field forwards.

Now that's completely understandable. After all, it's naturally difficult for the public to really get a good glimpse inside the sausage factory. But one of the things I really wanted to talk to you about is just that: looking inside the sausage factory. So here's my question to you: why haven't you written a popular book?

NAH: Well, people ask me this question a fair amount. Like most physicists, I really love talking about physics. I definitely enjoy giving public talks. I enjoy interacting with non-physicists about what's going on in the subject. It's something I've thought about, but I'm quite certain that I'm not going to do it for a very long time.

HB: Meaning what, exactly? Until you're old enough that you can't contribute much to science anymore?

NAH: Well, this is really the main point. There are many reasons why I'm not investing the time right now in writing a popular book, but the most dominant reason is that, while it's very important for the general public to know what's going on, and I definitely enjoy doing this sort of thing, it is not the most important thing for me to focus on right now.

I think that our real job is to push physics forward and to try to learn something new about the way nature works. And that's a very tough business. It sounds obvious, but it really is a very tough business. There are some people who, by whatever combination of their personal history and their talent or whatever, are positioned (or have already made significant contributions to) our understanding of the world.

And by this I mean something in a really serious sense. Not, *Are you one of the leaders of your generation?* or *Do you have a great academic job?* or something like that. I'm talking about things that will actually matter on the 100, 200, 300-year timescale, if not longer.

There are those who, by a combination of talent and luck and whatever else, have either done that already or are more easily capable of doing that. Then there are people who will never do that —and naturally not everyone in academia is focused in that direction anyway.

And then there are people who are right on the bubble. And I consider myself to be one of those. It's conceivable that I might be able to have some really important impact and push physics forward, but it's by no means obvious.

So my thinking is that the only thing that's in my control is giving my utmost to single-mindedly focus on the hardest problem, the most important problem that I have any hope of making some small progress on. The ability to concentrate and focus like that is absolutely crucial.

Of course it doesn't mean that I don't do anything else with my life. And actually, giving public talks about physics is one of the things that relaxes me. I enjoy doing it. It's an enormous amount of fun.

But I don't think of it even remotely as my actual job. My actual job is to try to figure out something about the way the world works. And I've been both blessed and cursed, I think, with just the right amount of talent, ability, motivation to have a chance of doing it.

HB: Why cursed, exactly? Because you'd like ten times more talent?

NAH: Well, ten times more would be fantastic. Ten times less it would be fairly clear that I couldn't do it, which would lead to a very different life.

Look, I realize that I'm in a phenomenal situation. But it means that the aspect of this business that is just flat out hard work is very important to me. It's the one thing that I feel is really in my control. If I didn't do that I would kick myself forever for not having given everything I had to try to do the things that are really important.

And for me everything flows from there. Our real job should be to figure out important things about the way the world works. We have huge problems: I mean really dramatic, zeroth-order, very important mysteries about the way the world works—things that anyone would be interested in knowing the answer to. Of course we're interested in knowing too. But my point is that they're not questions only of interest to egghead specialists: I think they're things that really matter to everyone. And we have the chance to tackle them.

It's not up to us when big breakthroughs happen; often big discoveries are made when it's their time to be made. And you could be unfortunate enough to live through one of the doldrum periods where what's going on is more or less an incremental addition to our general knowledge.

HB: And there are an awful lot of those.

NAH: Yes, there are an awful lot of those. But I genuinely have a sense —and I don't think it's just blind optimism, I really have this sense—that we're in a very exciting time right now.

After 300–400 years of barrelling through developments in physics that have given us some deep understanding about all sorts of basic things about the world around us, we're now at the point of addressing some of these very profound questions about where the universe came from, what's the origin of space and time, and so forth. These are finally the questions that are on the docket.

HB: I remember you once telling me years ago that you thought about this question of timing back when you were a graduate student, or perhaps even an undergraduate: that you'd been thinking about the timing of a major experiment like CERN's LHC in terms of how it might impact your career.

NAH: Absolutely. The prospect of all these wonderful experiments happening (even though they've been delayed a little bit) was something I was long anticipating, and played a major role in pushing me in a particular direction. You have to think of your research career on a big scale and over the long term. We only have thirty or forty years or so in which to try to push things forward.

So that's my view. I don't feel that this is one of these random times in the development of the history of the subject. I think it's conceivable that really, really big things are at stake, and making sure that we get that right is by far the most important thing that we can do as scientists.

Once we figure that out, telling people about it—telling people about whatever new-world view emerges if we get there—that's going to be incredibly important too. But it's not comparable in import to pushing things forward.

II. The Problem with Popularization

Not what it used to be

HB: So I get that. I get the fact that it's an important time; and I get the fact that, either way, it's *your* time, and you naturally want to be focusing on what you think are the most important issues and where you feel that you can make the most significant contribution. And I understand that writing a popular book would take a huge amount of time away from that. So, notwithstanding the fact that you believe passionately in the importance of communicating science and scientific ideas, you naturally want to focus on actually doing science.

But I'd like to move now in a somewhat different direction: are we actually doing a good job communicating science? Are the popular science books that are out there reflective of what's actually happening?

Suppose I'm some curious person who doesn't know much about science and wants to know what's happening at the front lines of research and am keen to buy a book about the Higgs boson or whatever. What sort of value am I getting? Am I getting someone's pet theory because he's pushing his agenda? Am I getting a distorted account?

NAH: This is obviously a really important question. It didn't use to be that there was such a divide between doing serious research and doing some kind of public outreach. There are all kinds of examples of absolute top theoretical physicists making the biggest breakthroughs of the time: Heisenberg, Schrödinger—

HB: Well, even going back to Faraday. Faraday used to give public lectures.

NAH: Yes, that's right. And George Gamow wrote these wonderful books. And then, famously, there was Feynman. I think if I had to name my two favourite popularizations of science it would be Feynman's Messenger Lectures at Cornell that turned into the book, *The Character of Physical Law*, and the book by Steven Weinberg, *Dreams of a Final Theory*. Both of these were written by fantastic theoretical physicists at the absolute peak of their powers in the midst of wonderful research careers.

HB: Who were not pushing their own agenda.

NAH: And that's precisely the point. I think that in popularizations of physics—I don't know about all the rest of the sciences to the same extent, but it's certainly true about popularizations of physics—there's just way too much about the latest, greatest idea: the latest trends in high energy physics, or whatever. I think that's just a big mistake.

There is not *remotely* enough about amazing things, about nature that we already know to be true and—even more importantly—the meta-issue about how we find out that they're true: how the scientific method actually works—not in this ridiculous, cookie-cutter, "hypothesis-experiment-back-to-hypothesis" stereotype that is so far from how things are actually done, but the actual story of how scientific discoveries are made and how wrong hypotheses are arrived at and eventually discarded. These are rollicking, fantastic, exciting stories. And we don't hear them. Well, at least, we don't see many of them in most popular books today.

Questions for Discussion:

1. What are the books by George Gamow that Nima refers to? Have you read any of them?

2. What do you think Nima means when he talks about the "ridiculous cookie-cutter" stereotype of the scientific method?

III. In Feynman's Footsteps

A genuine challenge

NAH: I think Feynman is a good example of how physics communication should be done. Those Messenger Lectures that he gave at Cornell were not even about the latest, greatest thing going on in particle physics in 1964.

HB: That he was so significantly contributing to.

NAH: Exactly—which he, together with other people, was busy revolutionizing, as he'd already done with QED (quantum electrodynamics). It wasn't even about QED, in fact. It was simply about what it sounded like: the character of physical law. He spent an entire lecture talking about Newton's Law of Gravity as an example of what a physical law looks like. He had five or six lectures all about basic things about the way the world works.

And it's amazing how much really deep stuff you can cover. I still find things in it—even me, a working theoretical physicist—that are inspiring, deep and important ideas about how we should actually go about doing this business.

HB: I notice that it's one of the few books you have on your bookshelves. I was shocked by how few books you have, but then you explained to me that that's because you're focused on new stuff...

NAH: Yes, right. But that's definitely one of them. And the other very inspiring thing about Feynman is that he was completely honest in how he discussed physics. I was very, very honoured to be asked to give the Messenger Lectures at Cornell back in 2010. And the principal reason I accepted is because Feynman's lectures there resulted in (tied with Weinberg's *Dreams of a Final Theory*) my favourite popularization of physics.

And I decided that I would try to do what he did *not* do, which is to talk about the future, to talk about what the current big problems are. I had five lectures, up to an hour and a half each. I spent the first ones talking about how we got here, and the last few talking about where we're going.

In trying to keep to its spirit I tried to be non-metaphorical and honest as much as possible, actually explaining what is going on by appealing to the underlying concepts—not some caricature of the truth but as close as possible to the actual truth.

Going in I thought it would be a piece of cake. Because what I spend a lot of my time doing in my work, like most of my colleagues, is to try to make progress on hard problems by understanding the simpler things as cleanly and perfectly and beautifully as possible. So all of the elementary things are in your head in some nice crystalline castle that you're happy to share with the world.

Or so I thought.

And I also thought it would be easy because I had so much time. I had given lots of public lectures before for an hour or an hour and a half; and because I had so much time on that occasion, it seemed that it would be really easy to explain it all.

But it was actually very difficult. I had a really, really, tough time of it. Overall I thought I did a reasonably good job, but in retrospect I certainly think I could have done some things better.

HB: Like what?

NAH: Well, I think everything could have been done a little more sharply. The discussion on cosmology could have been a little better. Some of the discussions of how we know what we know about quantum field theory were a little too metaphorical and fell back on a few too many of the standard analogies and metaphors that aren't really that good.

HB: But that's really your point, I suppose. It's really difficult to do.

NAH: Exactly. And even when I had a lot more time than most people have. Now admittedly it was an ambitious topic, but—

HB: But less ambitious than some of these popular books actually, when you think about it.

NAH: Yes, absolutely. So that's really one of the things that convinced me how difficult it was to do this. I spent two weeks of my life thinking about it really hard and planning it out, and not doing much else other than trying to do it right, and I think I did a reasonably good job. But I really learned how hard it is.

Feynman had to use all of his talent to the maximum to explain things like Newton's Law and the Second Law of Thermodynamics; and there are many more concepts even more abstract still that we need to get through now. I think trying to do it in a way that's honest really would be a full-time job for a significant period of time. It's not something I think that you can just do with your left hand. If you're right handed, that is.

But getting back to your question, that is what I think I would like to see more of. I'd like to see a lot more of telling people about what we actually know to be true about the way the world works.

There have been a couple of popular books in the past number of years a little bit along these lines. There was a very nice book about the Big Bang by Simon Singh. I thought that was an excellent book.

HB: Well, it's interesting because, of course, Simon Singh, while he has a particle physics background, is not a practicing physicist. So he's able to have that level of objectivity and that sense of, *This is an interesting story, I'm going to tell it*, as opposed to, *This is my theory or this is my agenda*.

NAH: Right.

Questions for Discussion:

1. Have you read Feynman's The Character of Physical Law or watched a video of his Messenger Lectures? Has reading this chapter impacted your motivation for doing so?

2. To what extent is it possible for frontline researchers to be both actively practicing science and responsibly communicating it? How does this difficulty influence the type of scientific understanding available to the general public?

IV. Describing Reality

The latest thing vs. the eternally significant

HB: Of course, it's also very understandable for a member of the general public to want to know what's new. *What's new in politics? What's new in biology? What's new in physics?* There's a natural curiosity there. But there are two essential problems with this as I see it.

In the first case, as you've alluded to, understanding what's new requires some basic comprehension of what's been already established. Which is why we find that all these books come out by all these different people, and the first half of them are all the same.

NAH: Exactly.

HB: It would be as if you'd want to buy a new work of fiction and the first half would be *The Brothers Karamazov* every single time.

NAH: Ha! Good choice.

HB: And then the second issue is that, as we've said already, there's a natural scepticism about why many of these people are writing their book to begin with. Again, at the end of the day it often boils down to, *"This is my theory"* or *"This is my interpretation."*

Now, on the one hand, that's completely reasonable. These people believe strongly in their work, they've often devoted the majority of their lives to it, so they want to tell you about what they're most passionate about and why. But on the other hand, if you really want to know what's new, if you really want to know what's cutting-edge, to some extent they're the *last* people you should talk to. You should talk to somebody with a much greater sense of objectivity.

NAH: So there are a number of aspects of what you're saying that are interesting.

One of the problems with talking about the latest, greatest thing, or whatever is going on at the cutting-edge of the field, is that it's changing rapidly. And when it's changing rapidly, which is a defining characteristic of any field in tumult, the principles that you think are foundational will later come to be seen as derivative. Things often get turned upside down because the right way of thinking about things has not yet been arrived at.

And this means that very successful popular books that provide snapshots of this process, can often 'freeze in' what are later seen to be the wrong pictures in the public consciousness. And that can be pretty counterproductive, because you end up spending a lot of time in later years undoing the things that were incorrectly laid down at an earlier time.

This happens a lot. I'll give you two examples just from theoretical physics. One of them is string theory.

Brian Greene wrote this incredibly famous and incredibly well-written book on string theory, *The Elegant Universe*. Unfortunately, from my point of view, it was written at exactly the wrong time and in exactly the wrong way. Because it was written smack in the middle of the famous "Second String Revolution".

And the most important thing that came out of the Second String Revolution was a *very* different picture than what people thought in the mid '80s—that particles aren't points, but little loops of string that move around, and instead of drawing these diagrams with points banging into each other you draw diagrams of little tubes interacting looking like pairs of pants, and talking about vibrating loops of string, and so forth. That was the cutting-edge point of view of the field back in 1985.

But by 1997 and 1998, the one thing that we knew about string theory was that it was **not** just a theory of strings. There were all sorts of other objects in it, and strings are just one aspect of it. There's something much, much deeper underneath it: there are these incredible dualities that illustrate that they're different descriptions of exactly the same physics.

And in fact, the entire source of tension and drama in the 1985 way of talking about string theory—that we've moved from a picture of good, old-fashioned particles and their interactions as described by quantum field theory, and then there's this radical generalization to string theory—no longer holds true. We understood by the late 1990s that these pictures are not only not at odds with each other, but are actually different descriptions of exactly the same system. That's perhaps the deepest insight we've had in theoretical physics in the last two decades.

And that insight was being developed, and everyone was talking about it, and it was the number one item on the agenda in theoretical physics *just as* this book was coming out promulgating the 1985 view that it's all loops of string and people playing violins and cellos and all of that.

And it just drives me nuts, because it's completely wrong. It's totally wrong. Of course, when I say totally wrong, I should add that people still call it string theory. Strings play a very important role—

HB: But it missed the big point at the time.

NAH: Exactly. There was this humongous point that was emerging, which will be remembered in the 100-year timescale. **This** was one of the things that will be remembered on a 100-year timescale.

It's a culmination, from my point of view, of 20th-century physics. We have these two big developments that are handed to us: these two revolutions in the first part of the 20th century concerning relativity and quantum mechanics.

And the two peaks of these revolutions are general relativity's representation of space, time and gravity and a relativistic understanding of quantum mechanics through quantum field theory.

But these two pillars seem to be fighting each other. Then at the end of the 20th century we realize that they're actually different descriptions of the same thing. This is an incredible fact.

Now, of course, Brian's a very smart guy; he knows this. He's written a whole series of books afterwards that increasingly keep backing into these things that we've learned. But in a sense it's too late. Now we have more or less hardwired into the public perception this old-fashioned picture that it's all about these little loops and not points.

But *that's* not what will be remembered. That's not eternally significant, the way that this amazing equivalence between these different theories *is* actually eternally significant.

And there are more examples like this. Something that is getting more popular recently—partially because the Higgs has been discovered and particle physics is getting a lot more attention—is that there are more and more people trying to explain quantum field theory in an accessible way.

And once again my buttons are being pushed, because they're explaining a point of view about the subject which is now about forty years old and which is almost certainly not going to be the way we think about it in the future.

Again, we're in the middle of some radical re-understanding of what this thing is really about. And when I say in the middle, there have been lots of remarkable developments in our understanding of what quantum field theory really is; they go back twenty years. We're learning more and more about it from lots of different points of view.

But the one thing that is almost certainly **not** going to be the case is the story that, *"The big deal is that there are these different fields and there are these particles that are excitations of the field."*

How often have you heard that recently in discussions of the Higgs? And again, it's not wrong in the sense that that's one language, it's one way of talking about it, but it's—

HB: Isn't this really just an example of grabbing the easy metaphor, taking the easy way out?

NAH: Well, I think the two things that I'm talking about here are not as bad as taking an easy metaphor. There's also the easy metaphor, which is really bad. We can come to the really lousy things in a second.

These things aren't actually lousy. At the appropriate time—which is not now, it should be stressed—at the appropriate time they *were* correct, in that that's what the people at the cutting-edge actually thought was the most important thing about what was going on. But again, that's why these are frontier subjects: they change.

One of the deepest things that we know about physical laws, which Feynman emphasized in his Messenger Lectures that I was talking about earlier, is that they have lots of different descriptions: the same laws can be talked about from many, many different philosophical starting points.

Which one you emphasize can start off being a matter of taste, but it's actually important to discover all the different ways there are of talking about it, because of all those different ways, one of them is best suited to hooking on to the next level of description.

And we've learned the hard way not to over-invest in our interpretation of the words we wrap around our equations. The words we wrap around them change while the invariant content stays the same. Different words become possible later, and sometimes those different words are deeper and more revealing.

But, of course, in talking to the general public, you have nothing but the words. Even when the words are as honest as they can be, it's hard not to make them misleading; and also hard not to make them eventually wrong.

HB: So by taking what I improperly called just now the lazy metaphor, which is really in this case the out-of-date metaphor—the way of looking at things that is no longer appropriate—you're committing a more egregious sin than just invoking an older-fashioned way of looking at the world. You're actually doing a disservice to the entire process of how our understanding evolves.

NAH: Exactly.

Questions for Discussion:

1. Has this chapter changed your perspective on popular science books, such as The Elegant Universe?

2. Are there always different ways of talking about the same physical laws? If so, what accounts for that? If not, what separates the different cases?

V. A Timeless Community

Walking with Galileo, aided by Weinberg

NAH: The entire language, the whole concept, of being interested in the most recent and most exciting thing is wrong when it comes to science.

We're not doing anything different—in kind, in spirit or in substance and structure—than what Newton was doing, what Galileo was doing, what all of our predecessors have been doing all this time. It's one of the most wonderful things about working in this subject: there is a seamless connection to all of our incredible heroes going back hundreds of years.

And you don't feel like you're doing something different. Sure, the tools are different: we have computers, we have graduate students, and so forth. But the basic process and the essential points—the way you think, the way you work—are exactly the same.

So this idea that it's only exciting when you learn about the latest thing is just totally wrong. There's this incredible architecture that's been built up over centuries and all of its rooms are magnificent. All the pieces that have been constructed already are sitting there. They are these jewels of our understanding of how the world works.

It's like going to some fancy museum and insisting on going to the room that's under renovation in order to see what's going on. That's kind of crazy. There are all these fantastic and beautiful things sitting there that demand respect and attention.

The really important thing is the strength and robustness of the architecture, how nothing could be a little different without crashing the entire structure. *Those* are the things to marvel at. *Those* are the things to understand. And by this incessant focus on the part that is under renovation, you're highlighting the sideshow without making clear the type of intellectual foundations that it's being built on.

HB: That's terribly frustrating if you're somebody who passionately cares about science, about "*understanding the character of physical law*" to quote Mr. Feynman. Because you'd naturally like to shout this from the rooftops: you think this is the most important thing in the world.

And if maybe members of the general public—in fact, not just the general public, but that subset of the general public who is sufficiently interested to shell out $30 for the latest popular science book—are constantly going into this one room in the museum under renovation, you might well think to yourself, '*Gosh, if we can't transmit these vitally important concepts to those people, that's a pretty frustrating state of affairs.*'

NAH: Yes, I think that's a real missed opportunity. I think, generally speaking, there's too much condescension to the general public when it comes to science. I'm not saying this is uniformly true of people who spend a lot of time popularizing science, but as a general statement I think it has some truth to it. People aren't stupid, and people aren't lazy.

Another one of my favourite popular physics books, favourite because it's where I started learning cosmology myself as a graduate student, was another book by Steve Weinberg, *The First Three Minutes*, which is about the big bang.

It's a beautiful, short book; and it's another one of these things that masterfully explains things with perfect accuracy. He says in the introduction that this book is going to take work. Its target audience is something like a tough old economist, someone who really wants to figure out what's going on and is willing to sit there and sort through and sift through things and actively struggle with the material.

I think that there are lots of people like that out there who we really need to reach. They're the people who I think would benefit from understanding the general structure of how we work, and also something about the more specific things that we're doing lately.

That book was another example of talking about very established, standard (by that time), physics. But, of course, there are also standard graduate textbooks in cosmology, and when I was a graduate student I started picking them up. But I couldn't make head or tail of them. So I actually went back and read *The First Three Minutes*, and then I got it. I got the general picture, and I went on from there.

Questions for Discussion:

1. Do you find Nima's metaphor of "the room under renovation" compelling?

2. What, if anything, do Nima's comments imply about the notion of "science journalism"?

VI. Against Relativism

Science, culture, and truth with a capital "T"

HB: There's this whole question about the process of science and understanding the internal culture: how the work is actually done and what's actually happening inside "the sausage factory." That is also something that many non-specialists are very, very interested in.

When I set up these public lectures at Perimeter Institute years ago, there were all these regular attendees who roughly fit the profile you mentioned a moment ago—like the tough old economist—people who were intelligent, curious, and motivated to understand. They'd go out on a rainy Monday evening, or whenever it was, to hear someone talk about science.

And at some very basic level, a question that many of these people wanted answered was simply: What are these theoretical physicists actually *doing* all day? There's this big building, and there are these guys in it—well, what the heck are they doing?

NAH: Right. Well, I can speak for myself—my 12 or 13-year-old self—because that's exactly what I wanted to know. That's what I really desperately wanted to know, because I knew, roughly, that this was what I wanted to do with my life. But what do you actually *do*? OK, you pick up a piece of paper and you start scribbling on it. But what are you actually doing? How do you know that there's anything to it?

And yes, I found that as well. I found that in interacting with the general public, the more insight you give them into how things are actually done, the more excited they are. But "actually done" often means at a very sort of nitty-gritty mechanical level: what you actually do, how you go about doing it.

But before I go on, I'd like to come back and say one more thing. Because you said the word "culture", which just reminded me of one of the catch phrases which goes along with this new way of popularizing science, which, as we said, is not the way it was even thirty or forty years ago.

This is the kind of phrase you hear a lot: *Science is culture—it's part of the general culture of humanity.*

And of course this is true to some extent. But this is another one of these phrases that really bothers me, because there's a sort of vacuous sense in which it's true. Everything is part of culture. We're all part of culture.

HB: It's tautological: it's hard to think of something that isn't culture.

NAH: Exactly, exactly. It's hard to think of something that isn't culture. But there's also a profound sense in which it's false. And the profound sense in which it's false is that science is really one of those things that's *independent* of culture. This is an essential aspect of what we're doing in science. We're discovering things that are actually out there. We're after the truth with a capital "T". It doesn't matter if you're Japanese or Zulu or Korean or German or American—

HB: Or lived a hundred years ago or will live a hundred years from now—

NAH: Or live in Alpha Centauri. We're after things that are actually *independent* of culture. That's precisely the point: that they have this eternal sense to them. They have nothing to do with the ups and downs of not only our day-to-day culture, but also the ups and downs of who's fighting who, who's killing who, and so on. These are all blips on the kind of timescale that we think about. And they're also even blips in the relative scale of importance of truth that we're talking about. So I think it's actually demeaning to science to make it part of culture in that sense.

HB: So if I can interject, because I think there's another thing somewhat linked to this that has always frustrated me, which I suspect
you might find a resonance with, a general conviction that there are
always two sides to every story.

If I'm a journalist and I hear a politician say, *"Well, you know, I
wanted to get this piece of legislation through but those guys on the
other side stopped me,"* I'm automatically thinking: *Well, that's just
one perspective. There are two sides to every story.*

And of course in their world there really *are* two sides to every
story, because you're always dealing with these two distinct camps;
and while there may be some objective truth somewhere, we'll
never actually be able to determine what it is because we don't
have access to it. That's simply the world in which they live, and it's
understandable that they come to those sorts of conclusions and
react accordingly.

But the scientific world is different. And it's different in the sense
that, while of course scientists may make mistakes and many scientific theories eventually become modified and the old theory is overthrown, you can't just randomly overthrow something holus-bolus. If
you overthrow a theory you have to explain why the previous theory
worked so successfully for so long before under past circumstances.

Which means, in fact, that there *aren't* two sides to every story
in the standard way that most journalists, say, believe. If somebody
comes along and says, *"Well, I think X,"* and somebody else retorts,
"No, you're just completely wrong," there is no possible world where
you can move to their particular theory if it doesn't already explain
what's happened before, if it's not sufficiently comprehensive.

NAH: It is definitely true that there is a sense of right and wrong in
science. And the sense of right or wrong is, for many of the reasons
that you're saying and some others that maybe we can get to later,
even more nuanced and deeper than most people realize. But of
course there's the most obvious sense that in the end it's experiment
that decides. So we do have this arbiter that we all agree upon, and
there can be no fights or disagreements about that.

But even in areas like contemporary theoretical physics when some questions seem beyond the realm of experiment for a very long time, even in situations like that, we have well-defined notions of right and wrong and well-defined metrics for progress so that we can still agree on some notion of truth with a capital "T".

Questions for Discussion:

1. *Do you agree or disagree with Howard's comments that there are not always "two sides to every story" in science?*

2. *To what extent do you concur with Nima's view that "science is independent of culture"?*

VII. Strongly Constrained

The effect of combining relativity and quantum mechanics

NAH: This last point isn't true in all fields of science, but it is true in our part of science, in fundamental physics. And that's because we've already come to such an astonishingly good and powerful description of the way the world works. We don't know the theory of everything, but we were handed some general principles in the last century: the principles of relativity and quantum mechanics.

And it turns out that, even if we didn't look at the way the world looked outside—if we didn't look to see that the world has gravity, it has light, it has these other forces that we discovered and understood in the early 20th century—if we just took these general principles of relativity and quantum mechanics and tried to figure out what could the world look like, just mathematically, from those important physical principles and following them to their logical conclusion, it turns out to be astonishingly constrained, whatever the underlying theory is.

We don't know what it is. We even suspect that incredibly basic things like space and time have to break down when we get down to Planckian distances—minuscule distances, 17 orders of magnitude smaller than any distance that we'll be able to probe in our lifetimes. So there are clearly enormous things that we're missing.

But these two big general principles tell us that what the world looks like at sufficiently large distances is relevant to anything that we might be able to probe.

Which is why I think that the current period in the development of physics is so important and so exciting. On the one hand, we're building off a stupendous foundation. Relativity and quantum mechanics together almost completely dictate what the world around us can possibly look like.

It gives us a tiny menu of possible particles we can have out there, and the kind of interactions that they can have are essentially dictated by these two principles.

Almost all the uncertainties about what the actual world around us looks like—all the freedom—comes in a choice of a small menu of particles and a small menu of interaction strengths. And that's it. We can't have 12,000 different types of particles and all different kinds of crazy interactions. The world is forced to be built out of very simple building blocks that are glued together in very, very, simple ways. So that's an amazingly strong foundation, and it works. We know it's right.

This is something that's really beautifully explained in Steven Weinberg's writings: you start with these two principles and then there's this amazing feeling of inevitability about what things can then look like that is consistent with them. The fact that they reproduce what the world around us looks like in all its gross features is just absolutely astonishing.

And this inevitability is associated with these fascinating dual properties that our laws of physics have.

On the one hand, they're remarkably rigid and universal.

On the other hand, though, they're extremely fragile, by which I mean that we take these two basic principles—the principles of relativity and quantum mechanics—for granted. But if we were to modify either one of them even a tiny bit then everything falls apart.

So they're very fragile to modification, but once you take these foundational principles they're incredibly rigid in what you're actually allowed to get out of them.

It was only in mid 1970s, in fact, that these ideas were fully appreciated, only then that we've had such an all-encompassing, powerful framework to think about physics: to have good control of the things we understand and a very good characterization of things that we don't understand.

And that's what's so exciting: that the foundations are incredibly strong. On the other hand, the mysteries are humongous, and they tell us that something dramatic has got to give in some aspect of how these basic ideas work together.

I think that that's something that isn't generally appreciated—the rigidity. There seems to be some sense that theoretical physicists, unencumbered by data from experiment, are just out there inventing leprechauns and fairies and nymphs and dryads around every corner; and every crazy idea you have is something that you can just put out there.

Of course, they say, *We all agree that experiment decides, but until experiment decides—*

HB: Everything goes.

NAH: Right: Everything goes. Leprechauns and nymphs and dryads are all on the same footing.

And it just is not like that. The incredible rigidity that we have in our framework makes it *almost impossible* to come up with a new idea. It is very hard to modify things in any way without ruining everything.

HB: So, where does this sentiment come from, you think? Why is there this general lack of appreciation of the rigidity of the principles?

NAH: Honestly, I think it's because there's this basic fact of the science that is not explained. I mean, it's explained in Weinberg's books. I'm happy to walk around and hand *Dreams of a Final Theory* to anyone I meet and say—

HB: "*Read it.*"

NAH: "*This is my philosophy. Please read it. Your life will be improved.*"

HB: There's your popular book right there.

NAH: Exactly. My popular book would be one page: "Dear Reader, welcome to my book. Read Steve Weinberg's *Dreams of a Final Theory.*"

Questions for Discussion:

1. Is physics a more "principle-based" science than biology? If so, why do you think that is?

2. In your view are the concepts of rigidity and fragility that Nima emphasizes here given sufficient attention in popular science books? If not, why not?

VIII. In Search of a Formula

Predicting clicks and theoretical candidates

NAH: But I really think that's it. There is this extraordinary fact about the way the world works that is not sufficiently appreciated. And it's naturally related to other misunderstandings that can arise. I was talking a little while ago about the relatively new trend of people starting to discuss quantum field theory.

So how do people typically do it? They say, *"Oh, there are particles out there in the world but we think of them as excitations of a field."*

Just look at that, right there: *"We think of them as"* seems to open up a humongous zoo of possibilities: well, **you** might think of them that way, but maybe there's some other way of thinking about them. And sure, it's been successful, your theory agrees with experiment— or at least, that's what you tell us...

We always confuse the language with which we describe phenomena with the phenomena themselves. And that's one of the deep lessons we've learned ever since quantum mechanics first shoved this down our throats in a lesson that we've had to digest over and over: you shouldn't confuse the language with which you describe the phenomenon with the phenomenon itself.

And when you focus on the phenomenon itself, there aren't these fields, actually: these fields are all in our head. That's one of the annoying things about this way of talking about things. Ultimately, there isn't a photon field, there is simply a detector. It measures photons.

It goes: click, click, click, click. It doesn't wave around or anything. It just goes click, click, click, click as the little photons come in. Things are made out of particles. There are things you can ultimately do with them: you can bang them into each other and scatter them around. Some go in, some go out.

Ultimately, you want some formula to tell you when all the clicks are going to happen, the frequency with which these clicks happen.

The world is quantum mechanical so we can't predict exactly what happens next, but we can predict probabilities for various things coming out when we collide some set of particles going in.

So that's already one really important thing to stress: there is some concrete objective in the end. You have to be able to predict what happens, at least probabilistically.

If you bring together any collection of particles, you have to be able to predict consistently what happens when they come out.

And there are very stringent consistency conditions on this. The constraints are that they have to be compatible with the laws of relativity and they have to be compatible with the laws of quantum mechanics. Being compatible with the laws of relativity means that you have to get the same answer no matter what frame of reference this collision happens in. And being compatible with the laws of quantum mechanics means that all the probabilities have to add up to one.

You might think that these are simple constraints to satisfy, but they're not. They're enormously difficult.

One of the fascinating things is that it's relativity and quantum mechanics *together* that make the situation so constrained. You could easily imagine a world without relativity, and you could easily imagine a world without quantum mechanics. Either way things would be tremendously less constrained. It's the existence of both of them that makes things so complicated.

Let me put it another way. If I were God—or a sub-God, say—and I was given the principles of relativity and quantum mechanics by the actual God who told me, *"Now go, build a world,"* I'd respond, *"Sorry, can't do it. This is just impossible."*

They seem almost completely incompatible with each other. In the end there are very few consistent interactions and consistent pictures that you can have for the probabilities for things coming in and going out that are compatible with the principles of relativity and quantum mechanics.

And this has nothing to do with the language that we use to describe it.

People found one language with which to describe it back in the 1920s, 1930s and 1940s: that's the language where there are fields, and particles are excitations of the field, and so on.

But we've also known for fifty years that there's nothing deep or fundamental about those fields. In fact, one of the early things people understood fifty years ago, is that you might call it one field and I might call it another. And we'd agree on the actual processes, even though the underlying fields we chose to describe them are different. So there's nothing deep about this idea of fields.

But again, I'd like to stress that anyone, whether inside or outside the mainstream of theoretical physics, is invited to come and make a contribution to the furthering of our understanding of fundamental physics. We're trying to do something very concrete.

Let's look at the famous problem of putting gravity and quantum mechanics together. An awful lot of nonsense is said about this problem too. It's often said that gravity describes the world at large, while quantum mechanics describes the world in the small, and we don't know how to put them together. That's just completely wrong.

Of course we can say the words "gravity" and "quantum mechanics" in the same sentence. That would be insane if we couldn't do that. The problem is that we don't understand what's going on at incredibly short distances, not that we can't say the words "gravity" and "quantum mechanics" in the same sentence.

But here too, is another topic where many think that everything is ruled by fashion: some people like string theory, some people like loop quantum gravity, some people like something else; and since experiment isn't going to decide anything anytime soon, people are going to argue indefinitely: it's nymphs, dryads and leprechauns to your heart's content.

But again—it's not like that. There is a formula that all of us are after. Gravity's important, so we're interested in interactions of ordinary particles with gravitons. We'd be happy—we'd be very, very happy—with a formula of the following type: take ordinary particles—maybe gravitons—and collide them at very high energies and figure out what are the probabilities for things to come out. All we ask is that they're compatible with the principle of relativity and the principle of quantum mechanics. That's it.

Please: provide that formula.

Of course, it will be a little involved because you can send anything in that you want and anything out that you want, but there is a very sharp mathematical sense of right and wrong even without one experiment. And that's because we know, from many, many, past experiments, about relativity and quantum mechanics. And that the two of them together represent such a powerful constraint on how the world works.

That's a formula that's missing for actual quantum gravity in our actual universe. Come up with that formula which is compatible with relativity and quantum mechanics. If you can come up with it, you have a candidate for a theory.

HB: That's a test.

NAH: Right. That is a test, even without any experiments: just a theoretical test. And that is a theoretical test that string theory passed in its early days and that we've learned more and more and more about since. That's why people got so excited about it. And at the time it involved some modification of the standard rules and the standard pictures for what people thought these probabilities could look like, while still being compatible with the principles of relativity and quantum mechanics.

As I said earlier, it seemed like a much more radical departure in the 1980s than it does now. Now we see in fact, that it's just another aspect of this general union of these two great principles.

In other words, it's physics, there's no doubt it's physics: it's about physical questions. But you can also turn it into a mathematical problem with a yes or no answer. And that's not something we could have done 150 years ago, which is a reflection of our great understanding of the world now: that we can formulate questions with well-defined answers, and if we had those answers we'd know a lot more about the way the world works.

Questions for Discussion:

1. How do you think Nima's views on the non-existence of fields would be received by other physicists?

2. To what extent is it reasonable to regard a graviton as an "ordinary particle"?

IX. A Principled Example

The inevitability of the Higgs

NAH: Let me give you an interesting example that illustrates what I'm talking about here.

This has to do with the Higgs particle. Everyone was very excited when the Higgs was discovered. But any decent theoretical physicist knew that the Higgs had to be there, more or less where it was found, for a long time now.

I once bet a year's salary, as a joke, while talking to some reporter that the Higgs would be discovered, and it ended up in some article and you would not believe the number of crackpot e-mails I got taking me up on this bet. But I meant it: I was completely sure that the Higgs would be there.

The whole story of the Higgs is a really nice example of the power of these general principles and how things work.

I mentioned before that the general principles of relativity and quantum mechanics tell us broadly how the world looks around us. But if you really carry out this program to its end you discover more details about the types of particles allowed. I should say first that one of the important features that elementary particles have is they have some spin: they're like little tops that have some angular momentum.

And you discover the only consistent theories that you can construct can only have elementary interactions involving three of these particles—not much, much more complicated interactions involving seventeen of them or something like that. In the end you can really build all the things that matter out of these little stick figures of just three interactions.

Secondly, the only spins they can have are zero, one-half, one, three-halves and two. For very good reasons, they have to be multiples of a half, and you also can't have $^{17}/_2$ or $^{33}/_2$. It can only be this tiny menu: zero, one-half, one, three halves and two.

Spin two has got to be unique. That's associated with gravity. And what had been seen in nature—up until a few months ago now when the Higgs was discovered—were the particles of spin one-half, spin one and spin two. For example, electrons have spin one-half, photons have spin one, and gravitons have spin two.

But it turned out that while studying in detail some of the properties of some of the other particles—cousins of the photons called the W and the Z particles that are related to the weak interactions, which are ultimately related to radioactivity—people discovered quantum-mechanical difficulties.

Again, you imagine the following process: you take some particles, bang them into each other, and try to predict the probabilities for what happens next. If you do this exercise for electrons and photons, everything is fine. The probability of them scattering is actually a fairly small number: it's around one in a thousand or one in ten thousand.

If you do it with gravity, by the way, at very low energies where we can imagine scattering them in real life, these probabilities are really minuscule. And one way of phrasing the difficulty of putting quantum mechanics and gravity together is that these probabilities grow with energy. So when you get to very, very, high energies they start getting bigger than one, and then you have a paradox.

But it turns out that there is a similar problem when you scatter these W and Z particles. You can do the calculation, and while the probability starts off looking small, just like it is for photons and other things, it actually grows with energy and becomes big again, just at an energy that's a factor of five or six higher than energies that we've been to before. So that's very exciting. It means that something new has got to happen before this disaster happens.

So then you just enumerate the possibilities of what the new thing could be. The new thing would have to involve some new particle that would change the calculus of this argument. And you go through the list of possibilities: it can't be spin two, it can't be spin three-halves, it can't be spin one, it can't be spin one-half. There is a fairly simple way (with some caveats) of ruling them out, until you're left with spin-zero, as by far the most reasonable thing that it could be.

And that's the Higgs: this new spin-zero particle that's there to solve the "disease" associated with the weak interactions. And you know everything about it: you know everything about how it has to interact with these particles, other ordinary particles, everything—just from the condition of solving this disease. We know everything about it ahead of time.

Now this is interesting, because just from thinking about the general principles, we have this relatively rich menu: zero, one-half, one, three-halves and two for the possible spins. We'd only seen one-half, one and two. And now we have some purely theoretical problem—nothing about the data, nothing that has suddenly come out of an experiment—but merely some theoretical anomaly associated with the W and Z particles and the weak interactions. And the only way of solving it within our completely conservative theoretical toolbox is to say that we need this spin-zero guy.

Now from one point of view, it's a very brave proposal. No one had ever seen something like that before, right? We'd never seen a particle like that before.

HB: On the other hand, you could say that it's inevitable.

NAH: Exactly. And that's the beautiful thing: that the principles are constrained enough that we understand what the possibilities are but they go beyond what we've seen.

Questions for Discussion:

1. To what extent does Nima's perspective suggest an approach consistent with Mathematical Platonism?

2. Has reading this chapter made you more interested in reading other accounts about the discovery of the Higgs boson?

X. Supersymmetry

Platonic convictions

NAH: John Wheeler mentioned long ago that there are two major approaches when it comes to dealing with mysteries in physics: radical conservativism or conservative radicalism.

Conservative radicalism is a dumb kind of radicalism. It goes like this: *What we will now do is list all the principles that we have and start throwing them out one at a time. Look, we're so brave! We're questioning relativity—we don't care about the authority of Einstein. We're questioning quantum mechanics.*

As far as I'm concerned, this is a pretty stupid way of proceeding. Things have never really worked that way. And in any case, everyone knows that you can always list all the principles and start knocking them out one at a time.

But like I said, physics has never worked that way. In fact, we've almost never really lost principles. The last time we were really seriously wrong about things was perhaps with Ptolemy.

HB: That was seriously wrong.

NAH: Yes. In fact, you can even argue that a little bit. But it was basically wrong.

But, really, ever since then nothing has been seriously wrong. Everything we've understood has been replaced by something deeper, but even that newer understanding always breathes more life into what came before, always explains features of what came before even more deeply. So we've actually never really lost principles.

At any rate, that's conservative radicalism.

Radical conservatism is exactly the opposite: you take what you have and you push it to its very limits, which are often vastly removed from where you had started.

To take one example, quantum mechanics started off describing the hydrogen atom and now we're applying it way, way, far beyond its original domain. So you take it as your first job to figure out what you can actually do, given the actual laws. The really good news for us is that, as I've said before, it's extremely constrained. There's not much we can do.

But, amazingly, it leaves open possibilities that have not actually been seen in nature yet. So we're in a funny situation where nature could do certain things, and we've only seen it do a part of what it could actually do.

As I said, the Higgs was a great example where we could actually say, *Look, there's something nature can do, we know that it can do it from looking at the situation from the top down; and it looks like that will solve a problem if we let it do it.*

And that's exactly the way the world works. Of course, it's a tremendous triumph for our experimental colleagues, but it's also an enormous boost to this belief in principles, this belief in taking what you have and pushing it to its absolute limits.

People are very excited by supersymmetry, but it hasn't been detected so far at the LHC. Again, I find it very frustrating that in public discussions, supersymmetry is often portrayed as another one of these nymphs, dryads and leprechauns: some people like it, some people don't, and so forth.

But to my thinking, supersymmetry is part of nature. It's part of nature in the sense that it's the last thing that nature actually can do compatible with these principles of relativity and quantum mechanics that we haven't seen.

HB: But does that mean that nature **must** do it?

NAH: No, no, it doesn't mean nature must do it, just like it didn't mean that nature necessarily had to choose the Higgs as a possibility.

People had thought about crazier things, giving up the principles a lot more. But the main reason theorists are excited about supersymmetry is that it completes this menu of actual possibilities that is extremely constrained.

Supersymmetry is associated with that last spin that was missing in the picture I was telling you about. Before I had listed the only possibilities as zero, one half, one, three halves and two.

We had only seen one half, one and two. We've now seen the Higgs, which is zero. The one that's missing is three halves. And the only way it turns out for there to be such a particle consistent with these general principles is for it to be supersymmetric and have all the properties that we expect from supersymmetric theories.

So it's a completion. If we saw supersymmetry it would mean that nature realized all the ways that could be compatible with these deep principles—there aren't seventeen other things that we could have seen as well.

HB: OK. But notwithstanding all of this, at what point would you be prepared to admit that supersymmetry is not the case? What would it take for you, and would you be prepared to wager any portion of your salary on that?

NAH: This is a very easy question, because supersymmetry as a general symmetry—as an idea that's realized somewhere, at some energy scale—is something that I would be willing to bet many years of salary on.

On the other hand, supersymmetry as something that will show up at the LHC is not something that I would bet a year of salary on. I still think it's one of the most plausible things that might happen, but I certainly wouldn't say with certainty that it will manifest itself at that particular energy scale.

HB: So you think it exists—

NAH: I think it exists Platonically. All of these consistent possible worlds, consistent with these general principles of relativity and quantum mechanics, in my mind I think of them all as real. I think of them as existing somewhere.

They're the things that nature can do compatible with these deep principles. And it's important that we know it exists. It was a big breakthrough to realize that it exists. Once more, it's important to stress that there are things that we arrive at by thinking from general principles, not just because we are confronted by something from experiment.

But then there are things that, from a very deep fundamental point of view, are details, but details that matter tremendously for the relevance of some of these things to particle physics.

Ultimately, the question about supersymmetry being realized at the LHC is related to a bigger question of whether the physics that we'll see will reaffirm this idea that the parameters and concepts of nature aren't particularly special—they could be anything they want to be and the world wouldn't change so much dramatically around us—or that they're very, very finely adjusted, and if you change them a little bit the world around us would change in a dramatic way.

We don't know the answer to that question. And that's really what's on the chopping block at the LHC.

Twenty years ago everyone assumed that we wouldn't have these very finely adjusted parameters, and that we would find some kind of new physics—supersymmetry or something else—that would explain things, that would give us the so-called "natural theory."

Now that very notion is in doubt. We might not. We seem to see— at least possibly, in cosmology—another example of very, very fine adjustment to parameters, just to explain the basic question of why the universe is big and isn't explosively accelerating or curled up to a really minuscule size. So it could be that we learn that lesson from the LHC. We just don't know. That's why we're doing the experiment. That is a question that we don't know the answer to.

But whatever the answer to that question—unless something truly massively revolutionary happens that somehow throws away all the success we've had up to now, which is extremely unlikely— we're going to see something compatible with these basic principles.

That's the remarkable thing. There are so few things that we can actually do. We're in such a straitjacket. It's so hard to have any new ideas that aren't obviously wrong that if you do manage to find them, you follow them up, you take them seriously. It doesn't mean that they have to be realized in nature. But the menu of possibilities is much smaller than people might think. There's a very small number of things that could be going on.

And we can arrive at figuring out what that small number of choices is largely from thought, given these general principles.

Questions for Discussion:

1. Is the view that nature should physically realize all the possible ways it can behave consistent with its founding principles itself another principle?

2. What does Nima mean exactly when he refers to issues of "fine tuning" and "naturalness" in this chapter?

XI. Reacting Precipitously

The sad tale of the supposedly superluminal neutrinos

HB: Several years ago now there was a big uproar with the announcement of so-called "faster than light neutrinos"—

NAH: Ah, yes. This is the other subject that I'd wanted to mention.

HB: It seems to me that there are a couple of interesting things to talk about here. In the first case, there is the question of how to properly deal with the natural excitement of a scientifically-oriented member of the general public who is captivated by the potential of a completely unexpected and earth-shattering new experimental discovery.

And then there's the matter of how specific results should be conveyed to the public. Many have wryly remarked to me that there was a tremendous amount of attention given to the original announcement trumpeting a scientific revolution and far less to the rather more sobering disclosure that the initial news was wrong.

NAH: I think it's actually a fantastic illustration of both how good science can work, how crappy science works, and also, unfortunately, how it's the more chaotic, muddled, junky way of proceeding that permeates the public consciousness.

And almost all the correct things I'm about to tell you about this whole episode were not reported. Some people said them—not even quite as correctly as should have been said, I must say—but they did not get remotely as much press as all the really shoddy sensationalism. So let's think about what happened there.

This experimental collaboration (called "OPERA") reports one part in a hundred thousand deviation in the speed of neutrinos from the speed of light. So it sounds like a tiny effect, right? Neutrinos are going one part in a hundred thousand faster than light over this distance that they traversed.

So how is this reported? Here is this incredible thing, you can go faster than light, and the sceptical physicists were largely described as people refusing to challenge the orthodoxy of big old Al who was looking down on us, wagging his finger and telling us, *"Don't you dare go faster than light!"*

Now, first of all, here's a little sociological fact about the way physics works that might be of interest to a young person: finding something that's actually right and as truly sensational as this is the greatest thing that could ever happen to you. So nobody is motivated to suppress something like this—far, far from it.

In fact, many people in the past had not only imagined that relativity could be violated—that different particles could go faster than light—but even many good theoretical and experimental physicists had done a lot of work on it. There was a nice industry—a small but serious industry—of people who had responsibly gone about characterizing how you'd look for it, over a fairly long period of time.

It really started in the early to mid-1990s with a number of different groups. Alan Kostelecky leads a group of fine physicists at Indiana, and very famous physicists like Shelly Glashow and Sidney Coleman also worked on the subject in the mid-1990s.

And they took two tracks.

One was, *Let's just see. Suppose these effects are there, and let's try to parametrize them: let's try to figure out how we would describe them in some robust language that allows us to make experimental predictions and correlate different possible experiments with each other.* So that's one thing that people tried to do.

Another thing they tried to do was to see if there could exist any decent underlying theoretical structure that might accommodate it. I myself did some work on this around eight or so years ago. It's conceptually straightforward to parametrize these effects. That's not hard. But the really interesting thing is that whenever you try to get them from some sort of underlying consistent framework that doesn't badly mess up other things, you fail.

And sometimes you've got to ask relatively deep questions—ultimately questions about black holes and black hole horizons in terms of whether or not you can associate a notion of a temperature with them the way Hawking did in a way that's compatible with the Second Law of Thermodynamics—in order to investigate what ultimately obstructs finding sensible theories that would allow you to violate special relativity in this way. That was an interesting theoretical activity that made it clear why it was so hard to actually do this.

But meanwhile, independent of how that went, there were people parameterizing what these effects could be: very responsibly and diligently figuring out how you could violate relativity far before any of those contentious experiments were done.

So the reason why all of us were sure this result had to be wrong was not, therefore, because we had never entertained the possibility that Einstein might be wrong. **Exactly the opposite!** That's what's so frustrating about it. We had entertained it *so* well, we had thought about it *so* much, *so* responsibly, and in such detail, that we *knew* it was impossible to have an effect as humongous as what had been alleged.

One part in a hundred thousand might sound like a small effect, but all previous work on violations of special relativity showed results that were much more stringent: one part in 10^{10}, one part in 10^{15}, one part in 10^{20}—just way off from these relatively humongous-size effects that they were finding.

So that was a big source of frustration: we *knew* it had to be wrong. People assumed that we were convinced it was wrong because we didn't want to question these underlying principles. Well, that's true in the sense that we know that you give up a lot if you do it. But we're not ideologues: we prepare for the possibility and study matters in so much detail that we know it cannot possibly be right, compatible with all the other experimental results we've had all this time.

So what happened then? Why did no one tell them? Why weren't they more sceptical about their results? Well, it's just a sad story all around. There were some papers written by theorists who had not thought about this stuff deeply: they didn't know the literature very well.

Mostly they were theorists who had talked to these experimentalists and were trying to motivate them to do a measurement like this by saying, *"Hey, you might see an effect as big as one part in a hundred thousand and it's compatible with everything we know."*

But they didn't do their damn job, because it's **not** compatible with everything that is actually known.

So there was a vicious circle going on there as those theorists and experimentalists reinforced each other. And then it so happens that they do see an effect that's that big. And in these junky theory papers—very careless theory papers that motivated the experimentalists—the main experimental constraint that they talked about was what we could measure from the neutrinos arriving from the 1987 supernova. It turns out, in fact, that they arrived a little earlier than the light but that's not because they're going faster, it was due to another effect. Anyway, the point is that that effect was understood. So we knew that there was nothing there.

But that was the one effect that they talked about in this really crappy theory paper. And that turned into the thing that all the journalists, all the science reporters everywhere, focused on, saying, *"Well, maybe there was a constraint on this effect and we would have seen something from the supernova already."*

That was neither here nor there. That constraint was trivial to evade, that constraint had nothing to do with the reasons why people who knew what was going on were sceptical about the result. But none of the responsible things that had been done for some ten or fifteen years actually made it out.

It's a great example of how the flip side of this rigidity is fragility. And both rigidity and fragility have great predictive power in telling you what to expect.

In the case of the Higgs we could predict something positive: there had to be this spin-zero particle there ahead of time: it has to couple like this, it has to be seen in that channel in this way. The theory is invented by some ten or so people fifty years ago. It's such a natural idea. And fifty years later it shows up in an experiment exactly where it's supposed to be.

That is a remarkable success story, and we see that by following this course we can predict possible things. But it's also true there are some things that are impossible. Of course, strictly speaking all we ever traffic with in science is degrees of certainty, but still this is so obviously improbable it is, to all intents and purposes, impossible.

So we knew, in the case of the faster than light neutrinos—not because of ideology, not because of blind deference to Einstein, but because of actually solid work—that the effect couldn't be there.

And that story didn't get out at all.

Questions for Discussion:

1. Are you surprised that the "really crappy theory paper" Nima refers to here made it through a peer-review process?

2. Should all science journalists have science degrees?

XII. Tangled Pillars

The relationship between relativity and quantum theory

HB: So the next time that something like this happens—and there will be a next time, of course—what should be done differently? Let's suppose I'm the science editor of some newspaper. What should the message be to my reporters? What should the message be to members of the general public?

NAH: I don't think that there has to be some monolithic message. Again, I think that it's important for people to understand essential aspects of how science is actually done. No one is out to suppress rebellious ideas: exactly the opposite. If you can find some even moderately rebellious idea that has even a modicum of truth associated with it, that's the way you make your name in the subject.

So no one is out there trying to suppress rebellious ideas. But again, it needs to be emphasized that we're in this very, very tight straightjacket. We're not going to destroy this entire incredible structure that we built up over four centuries and has served us so well, unless there's a *really* good reason for it—and we're certainly not going to do so at the drop of a hat.

Almost always, experimental or theoretical challenges to the structure are bound to fail. You shouldn't be surprised that they fail. And people shouldn't take scepticism as evidence of turf protection. It's really evidence of the great fact that we have not just one or two things that happen to work, but this entire castle that we built over centuries that works.

So people who say things that sound like the standard conservative viewpoint aren't just a bunch of fuddy-duddies who simply can't be bothered starting over. It would be amazing if we really were as wrong as Ptolemy was wrong and we'd have to start all over, but I think that's just incredibly unlikely to be true.

We've gone through so many revolutions now that have had this amazing feature. Nothing could be stranger and more remarkable than going from classical mechanics to quantum mechanics. But even that transition, once you understand quantum mechanics you say: *"Aha, there are these features of classical mechanics that make sense now!"* because they are embedded in this bigger structure in a way that explains connections between different ways of thinking about classical mechanics that we didn't understand before. So it would be astonishing to me if the entire structure would somehow just get pulled out from under our feet.

Of course that doesn't mean that there aren't things we don't understand, or that we're not going to have big surprises. I actually think that one of the big things that should happen in the 21st century in our part of theoretical physics is that we have to understand where space-time comes from in a more fundamental sense. We may even have to understand deeper things about quantum mechanics.

Quantum mechanics is unlikely to be wrong, but there are situations where it doesn't tell us what to do: there are confusing questions in cosmology where we ultimately don't know how to apply quantum mechanics. It's not that it's wrong, it's just sort of impotent to tell us how to make sense of some questions. So it may need some kind of extension to situations like that. Potentially. I'm not sure about that one. I'm a lot more confident—I think most people are a lot more confident—about the emergence of space-time from something deeper.

But regardless, that's the sort of "next thing," the next really big set of questions that's going to dominate the 21st century.

I think that one of the things that we will hopefully understand, when we understand things deeper, is why it is that these two big ideas of relativity and quantum mechanics seem to both fight each other so much, on the one hand (which is why the world is so constrained), but on the other hand, they also go so wonderfully together. In a world that didn't have quantum mechanics it wouldn't be that difficult to imagine modifying relativity a little bit, having it look more or less like relativity at macroscopic distances but departing from it at very short distances. Similarly, in a world without relativity, it's really easy to modify quantum mechanics.

Each one buttresses the other in a very strange yet really remarkable way. So we take these laws handed down to us by our ancestors—by Einstein, Bohr, Heisenberg, Schrödinger and Dirac and all these people—and when we try to put them together we find that we get into this big jam. It's so hard to make consistent things, and we come up with these tiny menus of possibilities. It's incredibly fertile and powerful, but that's all we can do. It's mysterious: in the way that they're handed to us they look to be fighting each other so much, where behind the scenes they seem to be helping each other so much.

Questions for Discussion:

1. How do you think a sociologist of science would respond to Nima's views?

2. Is "knowing how to apply quantum mechanics in cosmology" an issue for physicists or philosophers of science? Or both?

XIII. The Pull of the Truth

Plunging in, in the right vicinity

HB: So given all of this, what specifically excites you the most right now? What are you most passionate about, and where are you moving in terms of your own research?

NAH: Well, it's naturally very exciting to be in the decade of experimental discovery now, there's no doubt about that. We have the LHC, and we have all these astrophysics and cosmology experiments. And the consequent pick-up in the pace of just having to deal with buzz from experiments—little clues and rumours here and there—is like nothing seen before in the previous decade.

So that's an obvious change that I'm definitely excited about. And we'll have some answers to some of the questions that we care about. That it is really important to know. The question of whether there is some sort of fine-tuning at work in ultimately explaining the length-scale associated with the weak interactions is one that we'll have, perhaps not a complete, definitive answer to, but I'll be willing to bet, ultimately, years of salary on the answer to that question, given what we hear from the LHC over the next 5 or 10 years. So that's all moving along.

But what has also excited me a lot over the past three or four years involves developing a kind of a general strategy for trying to attack some of these deeper questions that we're not likely to get any direct hint from experiment on, such as, *Where does space-time come from?* I don't know if the strategy is right. I don't know if it's the best strategy for attacking this question. But at least it seems like **a** strategy for it.

This sort of qualitative strategy got me very excited a number of years ago and I've been working concretely along these lines ever since. I'm a newcomer to this business relative to a lot of other people who've been doing great things for one or two decades in this general kind of direction. Anyway, let me try to describe what this strategy is.

So we're convinced that space-time doesn't exist, it has to be replaced by something else. Well, what do you actually do with that? You can't simply wake up in the morning and say, *"Today I will work on emergent space-time!"* and just sit around at your desk all day wondering aimlessly about it. It's like the answer to my twelve-year old question I mentioned earlier: you want to know what space-time really is, but you can't just keep writing down, *What is space-time? What is space-time?* on your piece of paper. You need something to actually **do**. You need some concrete way to proceed.

That is the most difficult part of this whole business: we know what the big problems are, and we know that there are philosophies that guide us. All those things are there and everyone has that, everyone knows that. The most difficult part is to find the little chink in the armour of the problem that's going to allow you to go in through the side and somehow start making some progress on it.

The thing that we've discovered, really ever since quantum mechanics, is that we have to somehow be led to the answer by the answer itself. We can't just flat-out jump to the final formulation.

Progress is incremental in the sense that there is truth sitting there. And the wonderful thing about truth is that it's a great attractor: all you have to do is to get somewhere in its vicinity. And you also have to make sure not to fight it. Those are the two important things. And then you'll eventually get sucked towards it.

Just parenthetically, I think that if more people realized this then we'd get many more people interested in doing physics, because having the truth out there as a friend, as something that you're looking for and trying to head towards, is a tremendous leveller of the playing field when it comes to talent, inclination, particular mathematical strengths, and all of that.

Having nature as a guide and a friend makes a humongous difference, because people have vastly different levels of talent. Of course, you have to be very dedicated but still, you could be a fast worker, slow worker, whatever—all of these things pale in comparison to being somewhere in the vicinity of the right answer.

It's like having this thing as a friend and it just keeps telling you what to do next. You just have to head towards it, figure out some way of getting there, and just keep at it, keep following the important questions and see the light through the thicket as you get closer and closer to it.

The few things I've done in my career that I think are worth even something small have very much had this feeling to them. We're not inventing things: there are things that are out there, and we're sort of wandering around. We have to be sensitive to their presence, but as soon as you know they're there, it doesn't much matter if you have a bulldozer or a machete, or even a little crappy knife to go through the thicket: you'll get there. But it's important to get in its vicinity.

So this is what I'm trying to say: you can't just stand outside the forest and announce: *"In there somewhere is this great fountain of truth."* You have to go there and get scraped up, go down the wrong road and say, *"No, in fact that wasn't actually a beacon, it was just someone's flashlight that they had dropped on their way to the beacon,"* and things like that.

Question for Discussion:

1. What do you think Nima means when he says that, "We have to somehow be led to the answer by the answer itself"?

XIV. Choosing a Better Description

Thinking your way into the future

NAH: So you have to have something to do: you have to have some concrete angle into the problem. The most obvious angle into difficult problems is experiment, but we don't have that for this question of where space-time comes from. We're not going to do experiments at the Planck scale any time soon.

So my thinking is to look at the past for a guide. We've been through a similar huge conceptual leap before, when we went from classical to quantum mechanics. It's hard to imagine a bigger leap than that: we thought the world was deterministic, but it turns out that it's not.

And I imagine what it would be like—I often give this example when I give a colloquium on the subject—if you're a classical physicist in the year 1790 or something. You're working away and using Newton's laws to solve technical details related to the orbits of planets and stuff like that. And suddenly you're visited in the middle of the night by the Ghost of Theorist Future, who announces, *"I have a message for you from the year 1930: determinism is gone."* And then he disappears into the night, as Ghosts of Theorist Future are wont to do.

Now obviously, you're very excited by this. What are you going to do with this information? How does it change what you're doing? How, specifically, are you going to incorporate this news?

You obviously want to head towards this deep important thing that's going to come centuries from now.

So you could do various things. You could say, *"Okay, determinism is gone so I'm going to take Newton's laws and I'm going to add random crappy terms to them that somehow break determinism."*

That's an example of conservative radicalism. You say: *"Aha, I'm not so hidebound to Newton's strict determinism—look how radical and brave I am! I'm going to add these random, crappy, non-deterministic terms!"*

But you're *very* unlikely to get anywhere doing that; and the right answer looks nothing like that. In fact, it's really hard to even conceive of what the right answer looks like from that starting point. You have to change your entire framework and talk about wave functions and Hilbert spaces and all this stuff. You're *never* going to get to that. So, is it hopeless?

You could say, *"Well, I need some clues from experiment,"* but you'd have to wait over a century or so before you could start getting some valuable insights from experiment.

Yet there **is** something that you can do, which is to say, *"Look, if there's really no determinism, then there has to be some way of talking about even the physics I have right now under my feet in a way that somehow doesn't put in determinism. It can't really be there, so there must be some way of talking about it that doesn't have determinism in it."*

And that's a very startling thing. Because now, you see, you're on a straight and narrow path because you're not trying to guess the right answer, you're not trying to just mess with things randomly—instead you're trying to take this clue to force you to think about what you have under your feet in a radically different way.

And it turns out there **is** another way of thinking about good, old-fashioned classical physics: the idea that uses the principle of least action. When considering a particle that goes from one point to another, we consider all possible paths and recognize that it takes the path that minimizes the average of the kinetic minus potential energy along that path.

As it happens, the people who discovered the principle of least action were really startled by it philosophically, precisely because it looked so unlike Newton's laws.

It didn't look deterministic, but, of course, it was just a rewriting in the end: it was equivalent to Newton's laws. This is one of

Feynman's examples when he talks about this remarkable fact that you can take one set of theories and talk about it in so many radically different ways. And it's important to learn all the ways of talking about it, because only some of them might be better suited to understanding the next level of reality.

In this case, we now understand why it is that the principle of least action exists. It's because of quantum mechanics! The world is actually quantum mechanical, it isn't deterministic. And, indeed, this basic philosophy is correct in the limit as you can ignore quantum mechanical effects: the quantum formulation of the physics reduces to this new way of thinking about classical physics, which is the principle of least action.

Of course, I'm not saying that's the way that things actually happened historically—

HB: Sure. But that doesn't matter. Because history also didn't involve this guy coming from the future—

NAH: Exactly, exactly. All I'm saying is that if all you knew was that you had to lose determinism, you're not going to guess the right answer. But in order to actually go to work and do something, you could say to yourself, *"I need to think about Newton's laws in a new way..."*

HB: And then come up with the principle of least action.

NAH: Yes: and then come up with the principle of least action

Questions for Discussion:

1. Is it possible for the world to be, at its core, both quantum mechanical and deterministic?

2. What is the principle of least action and to what extent is it the same sort of principle as what Nima describes earlier when discussing relativity and quantum mechanics?

3. To what extent can thought experiments provide real knowledge about the world on a par with "real experiments"? Readers with a particular interest in thought experiments are referred to two other Ideas Roadshow conversations: In Chapter 7 of **The Problems of Physics, Reconsidered,** *Nobel Laureate Tony Leggett invokes a time-travelling analogy to justify a possible future breakdown in quantum theory that resonates strongly with aspects of Nima's thinking in this chapter, while in Chapter 3 of* **Plato's Heaven: A User's Guide** *philosopher of science James Robert Brown cites a variety of examples of influential thought experiments in both physics and mathematics.*

XV. Beyond Space-Time

Mathematics to the rescue?

NAH: As it happens, I think we're in a very similar situation today. It's space-time. We don't have ghosts of theorist future but we have these thought experiments that tell us that space-time isn't really fundamental. So what can we do about it?

Well, my colleagues here at the Institute for Advanced Study have made a huge amount of progress with this idea of holography that gives us a first example of what this idea of emergent space-time might actually look like.

As I said before, I think this idea is the culmination of 20th century physics by showing that all the threads of 20th century physics are unified. But I think we really need 21st century ideas in order to deal with the next step. In particular, we have to understand where time comes from. And physics, if nothing else for hundreds of years, has been about describing how systems evolve in time. Losing time is not a small thing.

That's why you can't just come to your desk and say, *"OK, today I will work on emergent space-time."*

But the idea is to try to describe the physics under our feet: nothing new, nothing different coming out of the experiments, just actual quantum field theory—this union of relativity and quantum mechanics—to try to describe things in a way that doesn't put in those principles as the starting point, but somehow has them emerge.

So it's a two-step process. First you have to figure out how to do that. That would be like finding the principle of least action. Then, an even bigger creative leap would be to figure out how it's deformed somehow and go to the next level.

So that's a general philosophy. It says that there are clues to what comes next that are hiding in plain sight underneath our noses in the structure of physics that we know for a fact works. But we have to reinterpret it and understand it properly.

What's kind of extraordinary is that when you actually study what happens in good, old-fashioned quantum field theories, you find that it's been screaming out to be thought of in a different way.

Suppose you actually do one of these calculations that we're talking about. You scatter an electron and a photon: they come in and then an electron and photon come out.

How do we calculate things like that? Well, Feynman famously taught us how to do that. He taught us to draw these pictures in space-time: his Feynman diagrams. You draw lots and lots of different processes, and each one of those processes describes a picture of what happens in space-time. So it's tied firmly to space-time through the idea that interactions take place at points in space and time.

It's quantum mechanical, so you have to sum over all the possible ways it could happen. And that's it. Figuring that out made Feynman famous. Essentially what he did was figure out how to talk about quantum mechanics and relativity in a way where both principles were manifested simultaneously. People didn't know how to do that before, but thanks to Feynman we now know how to do that.

But when you start looking at slightly more complicated processes there's a humongous increase in complexity of what the actual diagrams look like. There are tons of diagrams, and the expressions are 20–100 pages long. But people discovered, by lots of tricks, that the final answers could fit on one line: that they were astonishingly simple. These hundred pages would collapse to just a few terms.

And that has been a hint. That's been a hint in completely standard physics that's been sitting under our noses for sixty years now. There must be some other way of thinking about things.

Some of the work that I've been doing with many, many collaborators over the past number of years, involves discovering some really remarkable new mathematical structures that seem to be sitting underneath this physics.

That is, there is a way of talking about the physics that doesn't put in space-time, doesn't put in quantum mechanics, and gets the answers out in a very direct way. And that way involves a whole host of really new interesting structures in mathematics that are even new to the mathematicians, in parts of mathematics that have never had anything to do with physics before. They're related to algebraic geometry, and number theory, and ultimately—not very directly, but ultimately—a sequence of conjectures and ideas that were developed in response to trying to make sense of the Riemann hypothesis.

And that's one of the really exciting things. For a long time we've been wondering why number theory, which is one of the deepest parts of mathematics, has never had anything to do with physics?

But there seems to be some really deep, subterranean connection there. And the reason it's been hidden is because we've always insisted on describing physics in a way that made space-time and quantum mechanics manifest, in our face.

Well, we're not there yet, but it's very clear that there are fantastic new structures there—again, they're not speculative: it's just talking about standard physics, but in a very different way.

I'm hopeful that it will go somewhere more generally, beyond the very special simple theories that we're starting with to try to understand in this way. But it's revealing something deep about the way nature actually works. I would be thrilled if we find some way of talking about all of standard physics, not just the sort of toy models that we're looking at, but all of standard physics in a way that doesn't put space-time in, doesn't put quantum mechanics in, and gets the answers out.

And ultimately, then, if we really understand it, we'll have the beginning of an understanding of a starting point from which we can see the emergence of space-time and quantum mechanics. In some simple situations, we really understand it now.

HB: What sort of models are these?

NAH: They're toy models in the sense that they're supersymmetric theories with maximal supersymmetry, and in a certain limit where you don't look at the most complicated quantum corrections, but only the leading quantum corrections.

But they're genuinely quantum mechanical, relativistic theories where you can formulate them from the beginning using totally different ideas, completely different ideas, and then discover, *Oh, my goodness! Look! It's local! Oh, my goodness! Look! The probabilities add up to one!*

So we're starting to see that. And I'm very excited about that particular line of investigation, but more generally I'm very optimistic about this general strategy: to mine what we have underneath our feet for theoretical data, and try to understand it as deeply as we can.

Throughout history, when really big conceptual steps were needed, this has paid off time and time again. Einstein, for example, wasn't focusing on the latest experimental result when he thought about relativity.

There were these two deep things: Galileo and the notion of relativity, and Maxwell's view that the laws of nature were local. And he struggled with making these two things fit. It's really a small modification of Galilean relativity to make it consistent with the principle of locality; and that's what led to special relativity.

Or take the equality of inertial mass and gravitational mass. There again, it was just sitting there—a feature of Newton's laws—and he decided to make a big deal of it and understand it in a deeper way.

And as I've just said, even though people don't normally talk about it like this, I think quantum mechanics was presaged in classical physics—very much so—sitting there in the existence of these other strange formulations. Really understanding deeply why they exist and where they came from was the clue to the fact that quantum mechanics actually existed.

So it's not a controversial statement to say that nature has very few good ideas and she recycles them in subtle and interesting ways, over and over again. And it's our job to understand how that works.

So this is, I think, a very concrete strategy to proceed. It's not speculative: there's right and wrong. You can figure out if you have a new whiz-bang way of calculating all these things and understanding them, and you'd better agree with what everyone else had done before.

There are constraints and tests on whether what you do is correct or not. And I think it's a sure-footed path in this rough direction of what I think is the truth in this case: that there really isn't space-time, but there probably is some extension of quantum mechanics, and we need to somehow get there in an interesting route.

HB: Well, I think this is a very clear explication not only of your beliefs, but also what you do all day...

NAH: Yes, indeed. Yes.

HB: So thanks very much for your time. It's been great.

NAH: Thanks a lot.

Questions for Discussion:

1. What might be some of the thought experiments Nima alludes to in this chapter to make him suspect that space-time isn't really fundamental?

2. What are the strengths and weaknesses of working with "toy models" in the pursuit of a deeper understanding of nature?

3. Has this conversation given you a different appreciation for frontline theoretical physics research than other books or articles you have read?

Ideas Roadshow Collections

Each Ideas Roadshow collection offers five separate books presented in an accessible and engaging format.

- *Conversations About Anthropology & Sociology*
- *Conversations About Astrophysics & Cosmology*
- *Conversations About Biology*
- *Conversations About History, Volume 1*
- *Conversations About History, Volume 2*
- *Conversations About History, Volume 3*
- *Conversations About Language & Culture*
- *Conversations About Law*
- *Conversations About Neuroscience*
- *Conversations About Philosophy, Volume 1*
- *Conversations About Philosophy, Volume 2*
- *Conversations About Physics, Volume 1*
- *Conversations About Physics, Volume 2*
- *Conversations About Politics*
- *Conversations About Psychology, Volume 1*
- *Conversations About Psychology, Volume 2*
- *Conversations About Religion*
- *Conversations About Social Psychology*
- *Conversations About The Environment*
- *Conversations About The History of Ideas*

All collections are available as hardcover, paperback and eBook.

www.ingramcontent.com/pod-product-compliance
Lightning Source LLC
LaVergne TN
LVHW092010050326
832904LV00002B/45